Atlas of Ocular Anatomy

Mohammad Wakeel Ansari
Ahmed Nadeem

Atlas of Ocular Anatomy

Springer

Mohammad Wakeel Ansari, MD
Des Plaines, IL
USA

Ahmed Nadeem, DO
Core Faculty, Emergency Medicine
Residency Program
Midwestern University
Downers Grove, IL
USA

and

Provident Hospital of Cook County
Cook County Health and Hospitals
Systems, Emergency Medicine
Chicago, IL
USA

ISBN 978-3-319-42780-5 ISBN 978-3-319-42781-2 (eBook)
DOI 10.1007/978-3-319-42781-2

Library of Congress Control Number: 2016945151

Printed on acid-free paper

This Springer imprint is published by Springer Nature
The registered company is Springer International Publishing AG Switzerland

We dedicate this book to our wives Asefa and Farozan.

—Mohammad Wakeel Ansari
—Ahmed Nadeem

Preface

Let me start with a joke from my medical school days. We had a tough professor of anatomy who once asked a student: "What is the normal weight of a salivary gland?" The student had a sense of humor and instantly retorted: "Sir, with the capsule or without?" That well illustrates the dilemma of anatomy most of us have to face in our medical school days! But my personal opinion is that a sound knowledge of correlative anatomy of an organ can be an asset for a busy clinician who can thus easily anticipate its clinical presentation in disease. This atlas is a humble effort to present such knowledge for the eye.

We start with some solid examples. We can easily divide the eye anatomy into some regions, the starting point being the bony socket, the orbits on either sides of the nose, in which the eyes are safely lodged. Pyramidal in shape, the orbits have to have an apex on the back and an open base in the front. They are a bony socket closed by the orbital septum. They have a limited physical space suited for their normal contents—the eyeball along with its nerves and vessels, which come from middle cranial fossa through its apertures called superior orbital fissure and optic foramen.

So any growing lesion inside the orbit will displace its normal contents in a direction opposite to the growth. This may result in an oblique or forward proptosis of the eye, creating a loss of parallelism of the visual axis. It may also cause double vision (diplopia) (a good example is mucocele of the frontal or ethmoid sinus). A very severe proptosis, such as occurs in thyroid orbitopathy, may cause exposure keratitis because of incomplete closure of lids. A glioma of the optic nerve will cause a forward proptosis of the eyes. The structures inside the orbit are prone to trauma (contusion), which can cause fractures of weak spots in the orbit. Since the medial wall of the orbit is the thinnest, it can be fractured in severe contusions, resulting in air entry into the orbit and the periorbital tissues called crepitus; this, in turn, may cause bleeding from the nose if the patient blows his or her nose. Another weak spot in the orbit is that part of orbital floor near the infraorbital groove that changes into the infraorbital canal lodging the infraorbital vessels and nerves. In very severe contusions of the orbital margin, a sudden gross rise of intraorbital pressure can cause fracture of the floor (called a blow-out fracture) with herniation (displacement) of the neighboring inferior oblique or inferior rectus muscle into the underlying maxillary sinus. Because of the involvement of the infraorbital vessels and nerves, this may result in anesthesia of the upper lip. Thus a knowledge of anatomy of the orbit is sure to make things easier. Severe unilateral headache may be brought about by frontal sinusitis because this sinus exists in the superomedial angle of the roof of the orbit and may even trigger migraines in vulnerable female patients. Similarly, a complicated case of ethmoiditis in children may cause orbital cellulitis because a papery thin bone separates the eye from the ethmoid sinus. A painful red swelling just below the inner canthus is likely to be brought about by acute dacryocystitis because the lacrimal sac is situated there. Similarly, because the palpebral part of the main lacrimal gland is situated in the lateral part of the roof of the orbit, its inflammation, called acute dacryoadenitis, may present as a painful red swelling in the lateral part of the upper lid. We know that the sclera is very thin at the insertions of the rectus muscles on its front and therefore is a common site of rupture of the globe in severe injuries. The retina is thinnest at the ora serrate and the fovea; therefore retinal holes occur there when a traumatic detachment of the retina takes place. The visual pathway starts from the orbital, canalicular, and intracranial parts of the optic nerve, the optic chiasma, and the optic tracts, which end in the nucleus of the lateral geniculate body; there a synapse occurs and a new neuron of the nucleus of the lateral geniculate body takes over. The axons of these new neurons are further continued as the optic radiation, which finally ends in the visual center in the occipital lobe. If the arrangement of nerve fibers in various part of the visual pathway is known, the type of field of vision defects can be easily anticipated. These typical defects can help in localizing the site of a lesion in the visual pathway.

An ischemic microvascular lesion of the capillaries supplying a cranial nerve (as occurs in longstanding hypertension or uncontrolled diabetes) may cause an isolated palsy of the third, fourth, or sixth cranial nerves. The blood supply of the canalicular and cranial parts of the optic nerve is more vulnerable to trauma (especially the macular fibers) because the capillary meshes are wider and fewer vessels supply more nerve fibres via the septa of the pial network. The chiasma, in its position mostly above the pituitary fossa, is likely to suffer in tumors of the pituitary gland causing bitemporal hemianopia, especially with colored objects. Because the pupillary fibers separate themselves from the distal part of the optic tract, it is easy to remember that any patient with a lesion above the level of the lateral geniculate body will have normal pupils although he or she may be totally blind (cerebral blindness).

If one keeps in view the arrangement of nerve fibers in various parts of the visual pathway, it is not difficult to anticipate the type of field defects that may occur. For example, lesions anterior to the chiasma will cause unilateral field defects, while those posterior to the chiasma will cause contralateral homonymous hemianopia because the nasal fibers in the chiasma cross to its opposite side before entering the optic tract. A lesion in the occipital region tends to cause identical defects in each field, whereas optic tract lesions tend to cause dissimilar homonymous field defects. Because of its dual vascular supply, lesions of the occipital cortex may not affect macular fibers (macular sparing).

Optic nerve swelling (papilloedema) occurs mostly in lesions of the proximal part of the optic nerve but can also be seen in cases of raised intracranial pressure and compression of the orbital part of the optic nerve. This is because its cranial subarachnoid space freely communicates with the subarachnoid spaces of the orbital and canalicular parts. The most common causes are cerebral tumors, abscesses, subdural hematomas, subarachnoid hemorrhage, meningitis, and encephalitis.

Optic Neuritis

The most common cause of optic neuritis is demyelinating disease; it may be retrobulbar with a normal disc, but a painful sudden unilateral loss of vision with lowering of color vision and contrast sensitivity may occur, mostly in females in the fourth decade. An afferent pupillary defect is always present, and movements of eye maybe painful because of a blending of the sheath of the optic nerve with the origin of some rectus muscles.

Optic Nerve Compression

In optic neuropathy, which is not explained by intraocular lesions, compression of the optic nerve should be suspected. Early imaging of the orbit by MRI or CT scan is a great help. Cerebrovascular disease and tumors are responsible for most optic radiation lesions, although any intraocular disease can be involved.

Cranial Palsies

Ischemia (as in diabetes or hypertension), intracranial aneurysm, head injury, and intracranial tumors are causes of third cranial nerve palsy, which produces ipsilateral dysfunction. Aneurysms are more common at the junction of the internal carotid and posterior communicating arteries. Pupillary signs are common in these compression cases because pupillary fibers are superficial, whereas in ischemic lesions the pupils are normal.

Palsy of the sixth cranial nerve does not have much localizing value because it has a 90° bend at the apex of the petrous part of the temporal bone. Trochlear nerve palsies may be congenital or acquired.

Shape of the Eyeball

For proper refraction to occur at the cornea, it is essential that the walls of the eyeball are tight enough so that atmospheric pressure cannot indent them. This is achieved by circulation of a fluid inside the eye (the aqueous humor) that is chiefly secreted in the posterior chamber, from where it travels to the anterior chamber via the pupil. Then it goes to the periphery of the anterior chamber, called the angle of the anterior chamber; this structure has microscopic outlet channels of the aqueous humor (i.e., the trabecular meshwork, the canal of Schlemm, and the aqueous veins and collecting trunks). These drain the aqueous to the episcleral veinous plexus. This circulation of the aqueous maintains a positive pressure inside the eye called the intraocular pressure, which is responsible for maintenance of the shape of the eyeball and is essential for refraction at the cornea. Because of this intraocular pressure, atmospheric pressure cannot indent the eyeball. For proper maintenance of intraocular pressure within normal limits (12–20 mm Hg), the anterior chamber should have an optimum width so that the aqueous has adequate access to the microscopic outlet channels situated there. A narrow angle and a shallow anterior chamber can cause a sudden gross rise of intraocular pressure up to 60 mm Hg, creating a sudden stretching of the corneal nerves and severe pain and vomiting, a condition called acute congestive glaucoma, which is a serious ophthalmic emergency. Therefore before dilating the pupils for examination of the fundus, one must make sure that the anterior chamber is not shallow and that its angle is not too narrow. This can be clinically tested by the Iris Shadow Test.

No book can claim to be perfect including this one but since our medical school days we have wanted to see an anatomy book that could be simple, self-explanatory, and at the same time enjoyable. This demanded a correlation of structure with function. This also demanded highly schematic diagrams instead of photographs of dissected specimens that could aid the easy grasp of the subject. This atlas is our humble effort to achieve this.

In the end, I must thank the team of Springer editors whose cooperation was an asset for me in preparation of this atlas. Special thanks are due to Dr. Asefa Ansari and Dr. Farozan Islam who helped me whenever I got stuck up with my computer.

Suggestions and comments are welcome.

Illinois, USA Mohammad Wakeel Ansari
Illinois, USA Ahmed Nadeem

Contents

Anatomy of the Orbit

1

Keywords
Four-walled orbital cavity • Bony socket for eyeballs • Pyramidal shape • Apex and base of orbit • Paranasal sinuses • Frontal ethmoidal • Maxillary • Roof • Floor • Lateral and medial walls of orbit • Mucocele • Proptosis

The eyeball is safely lodged in a bony socket called the orbit on the two sides of the nose. The four-sided orbital cavity is pyramidal in shape with an open base in front and an apex on the back. The four walls of the orbit converge posteriorly to reach the apex on their back formed by the optic foramen. The four walls are called the roof, floor, and medial and lateral walls. The lateral wall, which has a protective function, is the strongest. The medial wall is the thinnest and contains the ethmoidal sinuses with ethmoid air cells. There are three sinuses—anterior, middle, and posterior—on each side. They are liable to fracture with severe contusion injuries of the orbit, in which case air may enter the orbit and periorbital tissues, causing periorbital crepitus. Nasal bleeding may occur if the patient blows his or her nose. Another weak spot in the orbit is in its floor near the infraorbital groove. This changes into the infraorbital canal, which lodges the infraorbital nerves and vessels whose involvement may cause anesthesia of the upper lip in what is called a "blow-out fracture of the floor." Three paranasal sinuses are present in the orbit: (1) the frontal in the superomedial part of the roof, (2) the ethmoidal medially, and (3) the maxillary in the floor. In the blow-out fracture of the floor, herniation of the inferior oblique muscle and of some fat into the underlying maxillary sinus may occur. Diseases of these paranasal sinuses may affect the orbit secondarily, such as in orbital cellulitis in untreated ethmoiditis in children, and mucocele of the frontal sinus can cause downward and outward pushing of the eyeball. This is called oblique proptosis. Orbital space is limited to the orbit by the septum orbitale, and a growing mass will displace its normal contents in a direction opposite of the mass. A good example is mucocele of the frontoethmoidal sinus, which may displace the eyeball down and out.

1.1 Basic Concepts

The four-walled bony orbital cavities form a sort of protective socket for each eyeball and are situated on the two sides of nose (Figs. 1.1 and 1.2; Table 1.1). It is in this bony socket, in which each eyeball is lodged along with its muscles, nerves, and blood vessels that are essential for its

M.W. Ansari and A. Nadeem, *Atlas of Ocular Anatomy*, DOI 10.1007/978-3-319-42781-2_1

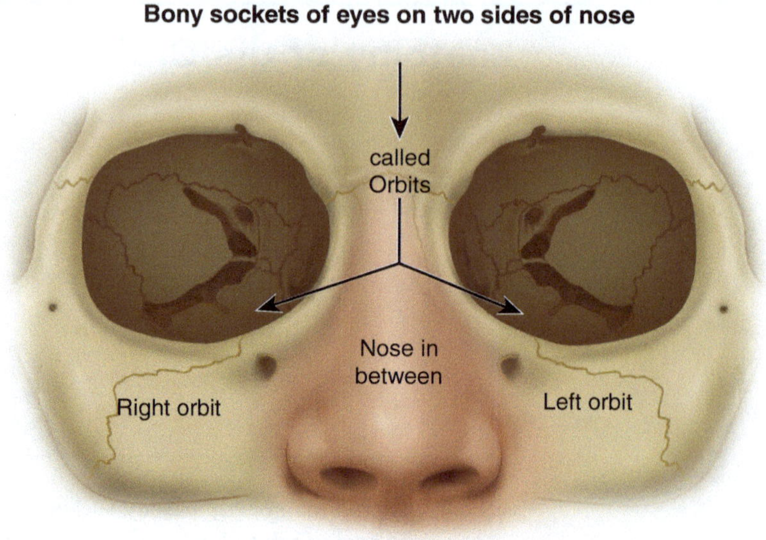

Fig. 1.1 The nose between two bony sockets (orbit)

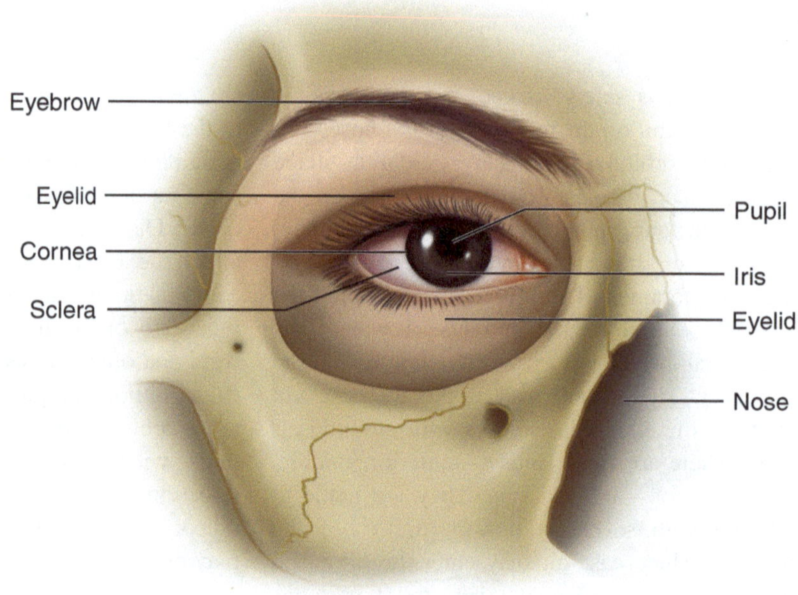

Fig. 1.2 The eyeball in its bony socket

proper functioning. Above each orbit is the anterior cranial fossa, medially are the nasal cavity and ethmoid air sinuses, and below is the maxillary sinus in its floor. The four walls are called the roof, the floor, and the lateral and medial walls. The cavity is pyramidal in shape with a quadrangular base in front and an apex behind. The four walls converge behind and appear to meet on the back, the apex, which is formed by the optic foramen (OF). The base is in front, formed by a four-sided orbital margin. The cavity is often compared to the shape of a pear whose stalk is formed by the optic canal. Its medial walls are parallel to each other (Fig. 1.3). The lateral wall makes an angle of 45° with its medial wall, and the axis of each orbit makes a

Table 1.1 Summary of the bony socket of the eyes, called the orbit

The roof has two bones: (1) The frontal and (2) Greater wings of sphenoid bone. It has four important structures medially: the frontal sinus, trochlear pulley, and trochlear fossa, and, laterally, the main lacrimal gland (Fig. 1.6)
The floor has three bones: (1) Orbital plate of maxilla, (2) Orbital surface of zygomatic, and (3) Orbital process of palatine bone. It has an infraorbital groove, which changes into infraorbital canal, causing anesthesia of the upper lip because of infraorbital nerve involvement (Fig. 1.7)
The strongest lateral wall (Fig. 1.8) has two bones: (1) Zygomatic and (2) Sphenoid. It includes the Whitnall's tubercle, which attaches to the following: (1) Lateral palpebral ligament (2) Lateral check ligament (3) Lateral edge of aponeurosis of levator palpebrae superioris
The thinnest medial wall (Fig. 1.9a) has four bones: (1) Frontal process of Maxilla (2) Lacrimal bone (3) Orbital plate of ethmoid bone (4) A small part of body of sphenoid bone
It has a lacrimal fossa between the anterior and posterior lacrimal crest, lodging the lacrimal sac, which opens into the nasolacrimal duct (NLD), opening into the inferior meatus of the nose

23° angle to the visual axis. The lateral wall is the strongest for protection and the medial wall is the thinnest. Only a papery bone separates the eye from the ethmoid air sinuses, which is why air from ethmoid air cells may enter the orbit and periorbital tissues, causing periorbital crepitus and bleeding from the nose in severe contusion injuries of the orbit.

1.2 Paranasal Sinuses and Gaps in the Orbit (Fig. 1.4)

There are three paranasal sinuses in the orbit: the frontal in the superomedial part of the roof, the ethmoid medially, and the maxillary sinus inferiorly in the floor. The OF is actually the orbital opening of a bony canal, which opens at its back into the middle cranial fossa of the skull, through which the optic nerve and ophthalmic artery pass. This canal communicates the cranial cavity to the orbital cavity, and its orbital opening is called the OF. There are three gaps in the orbit: the superior orbital fissure (SOF) separates the lateral wall from the roof; the inferior orbital fissure (IOF) separates the lateral wall from the floor; and the OF and is situated at the apex. There are four tubercles in the orbit (Fig. 1.5): the Whitnall's and zygomatic tubercles in the lateral wall, the lacrimal tubercle in the inferior orbital margin, and the infraoptic tubercle in the medial margin of superior orbital fissure. Attachments to these tubercles are elaborated in Fig. 1.5.

1.3 Details of the Roof, Floor, and Lateral and Medial Walls

Figure 1.6a, b shows the triangular roof of the orbit formed mainly by the frontal bone and a small part by the lesser wing of the sphenoid sinus. The apex of triangle is at the back near the OF. The base of triangle is the superior orbital margin. The roof has a fossa in its lateral part that lodges the main lacrimal gland. Figure 1.7 shows details of the floor formed by the orbital plate of the maxilla, zygomatic, and palatine bones. The infraorbital groove changes into the infraorbital canal in the floor, which opens into the infraorbital foramen (Fig. 1.7). This canal, which holds the infraorbital vessels and nerves, constitutes a weak spot in the orbit that gives way in blow-out fractures of the orbit caused by the forceful impact of a squash or tennis ball or a clenched fist on the orbital margins (Fig. 1.8). A sudden gross rise in intraorbital pressure will cause this fracture, creating anesthesia of the upper lip by involvement of infraorbital vessels and nerves. This severe contusion injury may cause a displacement (herniation) of neighboring inferior oblique muscles and a limitation of elevation of the eye with binocular diplopia. Figure 1.9a–c details the strongest lateral wall formed by the greater wing of the sphenoid and zygomatic bone. It is separated from the floor by the IOF.

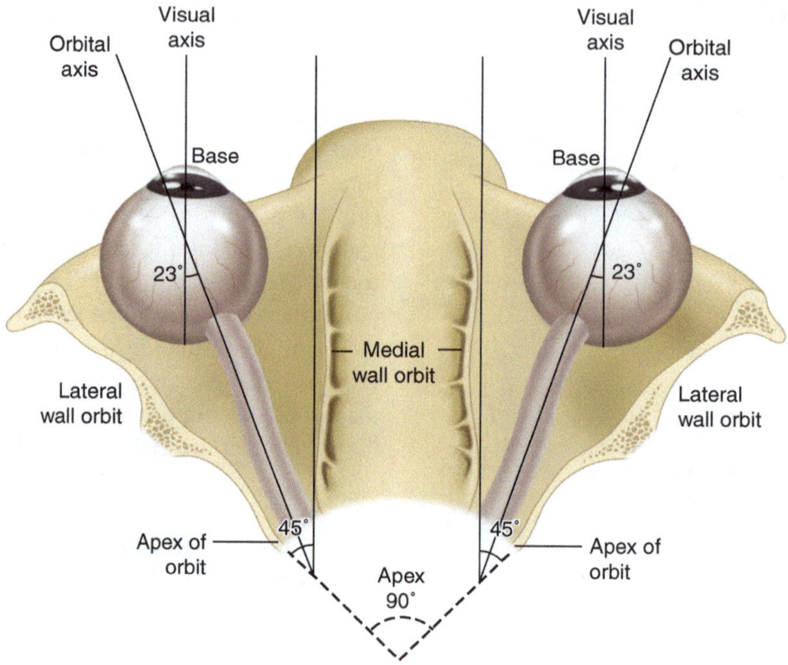

Fig. 1.3 Eyeball in pyramidal bony socket (orbits). Medial walls parallel with each other. Visual axis makes an angle of 23° with the orbital axis; lateral and medial walls make an angle of 45° with each other

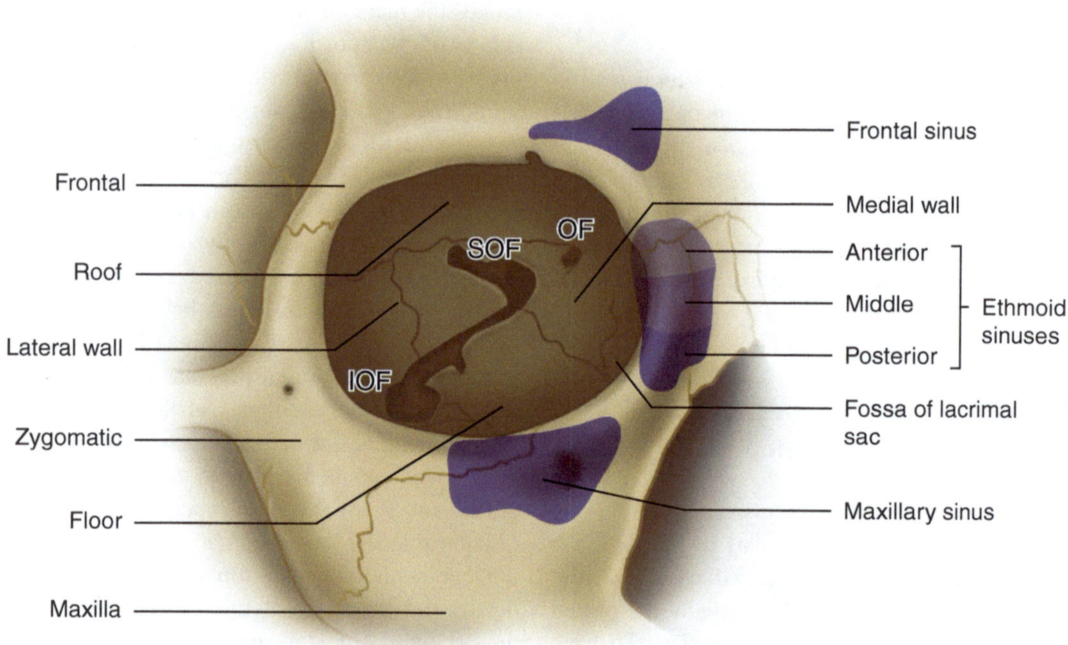

Fig. 1.4 Right orbit from front (*schematic*). Note three sinuses: (1) Frontal (superomedial); (2) Ethmoidal (medial); (3) Maxillary (in floor). Note three gaps: optic foramen (OF), superior orbital fissure (SOF), and inferior orbital fissure (IOF). Note the three bones in the orbital margin: (1) Frontal; (2) Zygomatic; (3) Maxillary

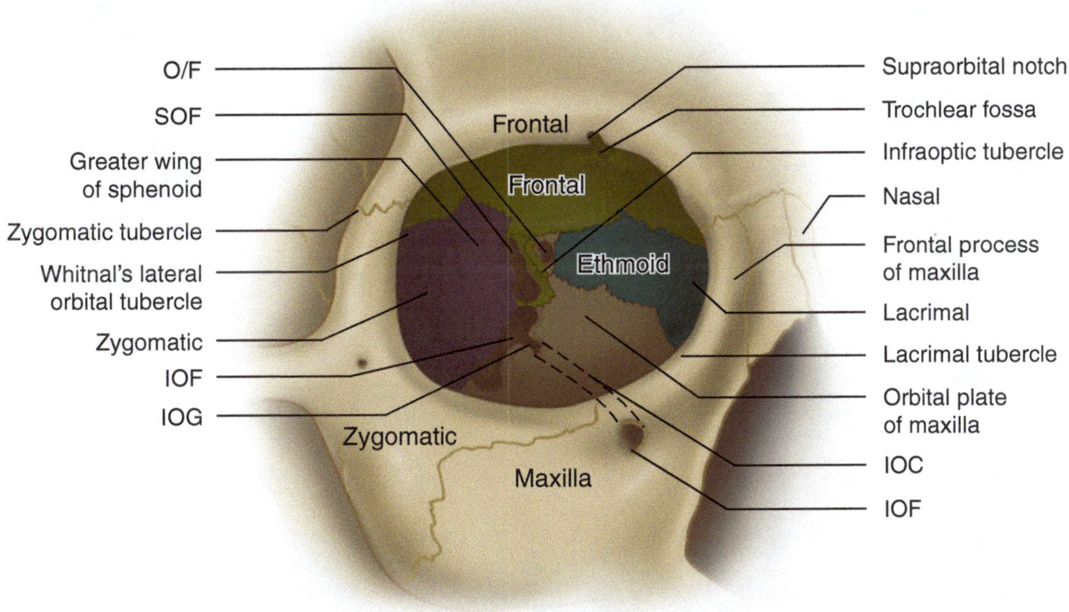

Base of orbit quadrangular in front
At apex orbit
On back – O/F - Optic foramen
SOF - Superior orbital fissure
IOF - Inferior orbital fissure
In floor orbit – IOG - Infraorbital groove
IOC - Infraorbital canal
IOF - Infraorbital foramen

Fig. 1.5 Four tubercles in orbit: (1) Whitnall's tubercle in lateral wall; (2) Zygomatic tubercle; (3) Infraoptic tubercle; (4) Lacrimal tubercle

1.4 Weak Spots in the Orbit

We have already mentioned that the infraorbital groove and canal constitute weak spots in the floor of the orbit. Another weak spot, as shown in Fig. 1.8a, is the thin bone ethmoid, which separates the eye from the medial wall. Ethmoiditis in children may give rise to orbital cellulitis, and severe contusion injuries can cause rupture of the medial wall and periorbital crepitus. Figure 1.9a also shows the lacrimal fossa in the medial wall. The fossa lies between the anterior and posterior lacrimal crests and lodges the lacrimal sac, which is continuous below with the nasolacrimal duct. The duct's lower end opens into the inferior meatus of the nose for drainage of tears when severe emotional stress occurs.

The orbital cavity is divided into a central and a peripheral space by a cone of muscles formed by four rectus muscles joined by the intermuscular septum, which extends from the apex of the

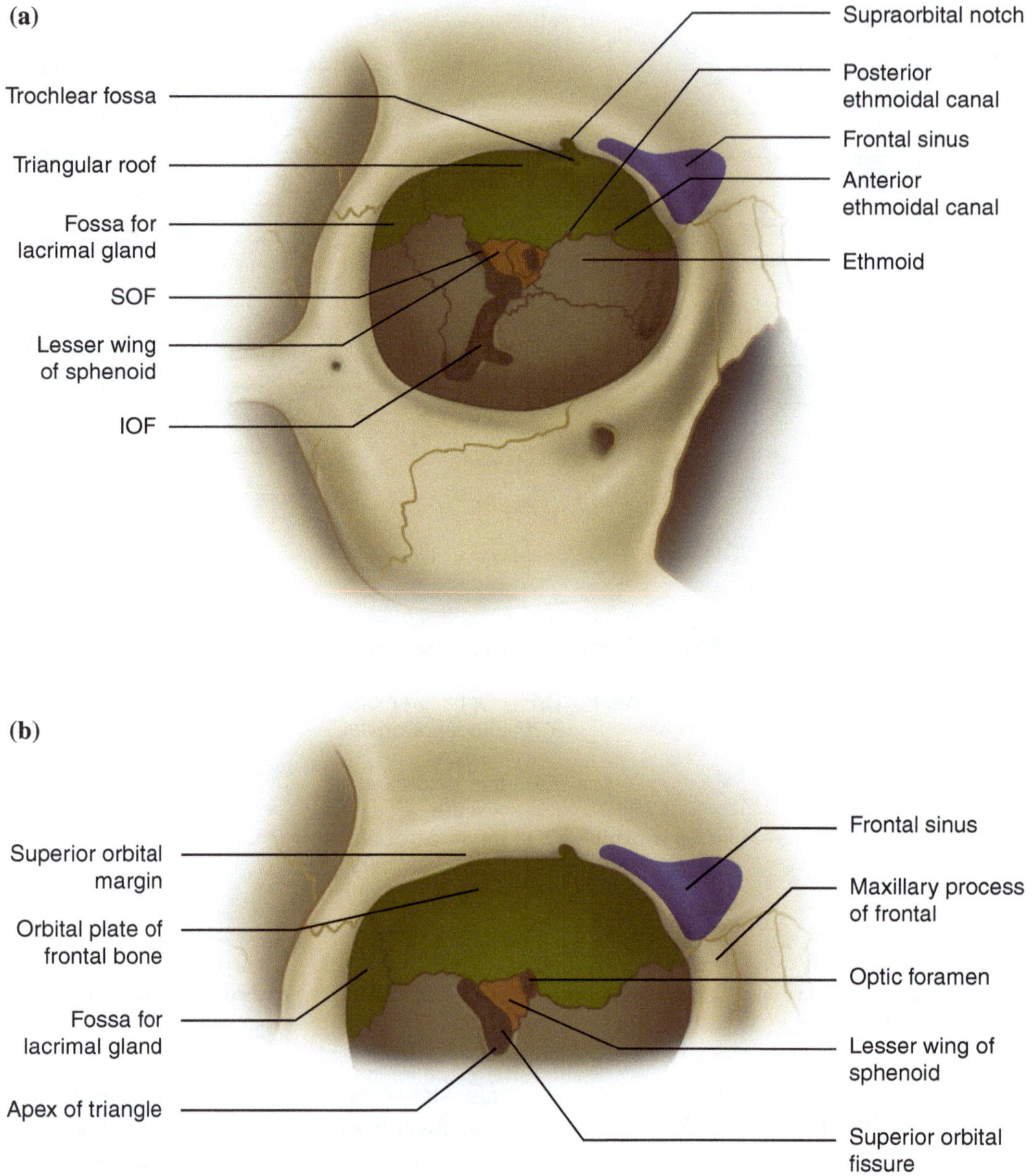

Fig. 1.6 **a** Roof of the orbit (two bones): (1) Frontal (major part); (2) Lesser wing sphenoid. **b** Triangular roof of orbit: (1) Apex on back lesser wing sphenoid; (2) Base: superior orbital margin

orbit to the eyeball. Inside the cone is the central space, which has a central pad of fat along with the optic nerve surrounded by vessels and nerves. The peripheral space has the peripheral pad of fat and the lacrimal nerve and the vessels supplying the structures outside the cone.

1.5 Orbital Margins

The superior orbital margin has a fascial sheet extending from the superior orbital margin to the upper border of the upper tarsus. Similarly, the

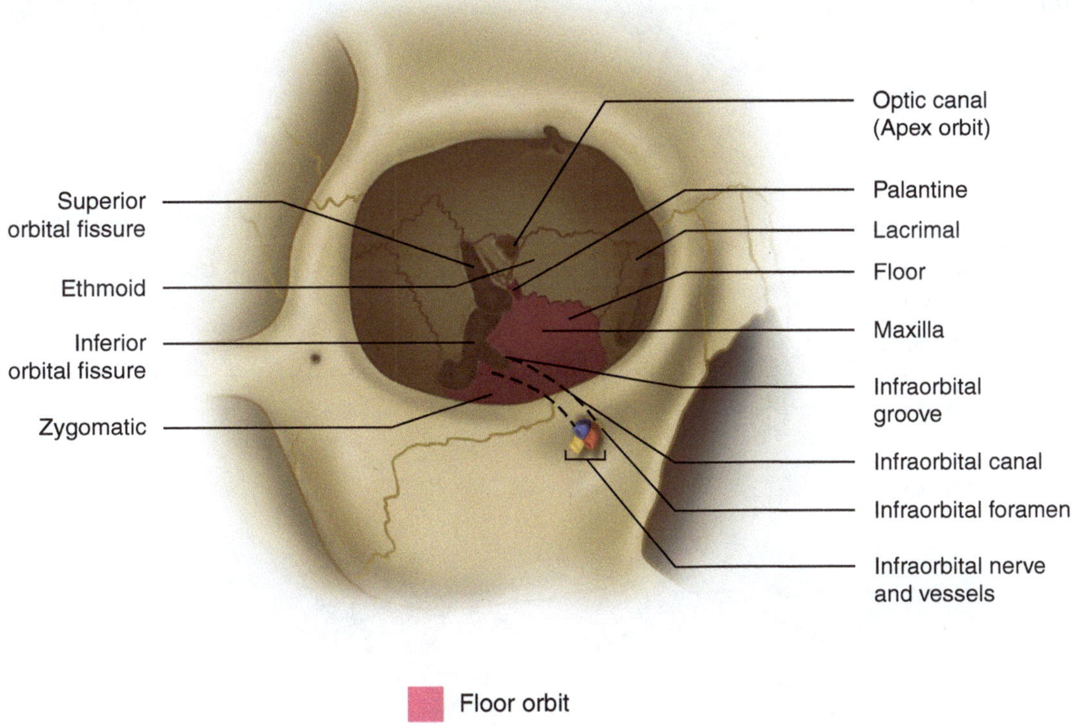

Fig. 1.7 Right orbit from front. Note infraorbital groove changing into infraorbital. Infraorbital canal in floor of the orbit forming the weak spot in blow-out fracture

inferior orbital margin has a sheet of fascia extending from the inferior orbital margin to the lower border of the lower tarsus, the back or harder part of the lower lid. These sheets are called septum orbitale; along with the tarsi of the upper and lower lids, they almost separate the orbital contents from the lids. That explains why an inflammation behind this septum (postseptal inflammation) may take longer to present clinically while a preseptal inflammation (e.g., blepharitis, lid abscess, stye) may not reach the orbit for some time.

The orbital cavity thus may have a space suitable for its normal contents (eyeball, nerves, muscles, and blood vessels), but if a growing mass develops inside the orbit, it will displace some of the orbital contents such as the eyeball itself, or the mucocele of the frontal or ethmoidal sinus may push the eyeball down and out, causing oblique proptosis and binocular diplopia because of a lack of parallelism of the two eyes. Tumors inside the cone of muscles such as a glioma of the optic nerve or a bleeding behind the eyeball (retrobulbar hemorrhage) cause a forward proptosis, and mucoceles of the frontoethmoidal sinuses cause an oblique proptosis. In these space-occupying lesions, a scleral rim may be seen in the primary position of the eyeball.

In very severe cases of proptosis such as thyroid orbitopathy, the lids may not close the eyeball properly, and exposure keratitis can occur if this is left unattended.

When the lids are closed, the conjunctival sac becomes a closed space. This conjunctival sac is indirectly connected to the mucus membrane of the nose through the lower opening of the nasolacrimal duct into the inferior meatus of the nose. Thus organisms affecting the nasal mucus membrane may secondarily affect the conjunctival epithelium, causing conjunctivitis. Since the ethmoid sinus is the closest to the eyes, untreated ethmoiditis in children may lead to orbital cellulitis. Frontal sinusitis can cause severe unilateral headache and may precipitate migraine

(a)

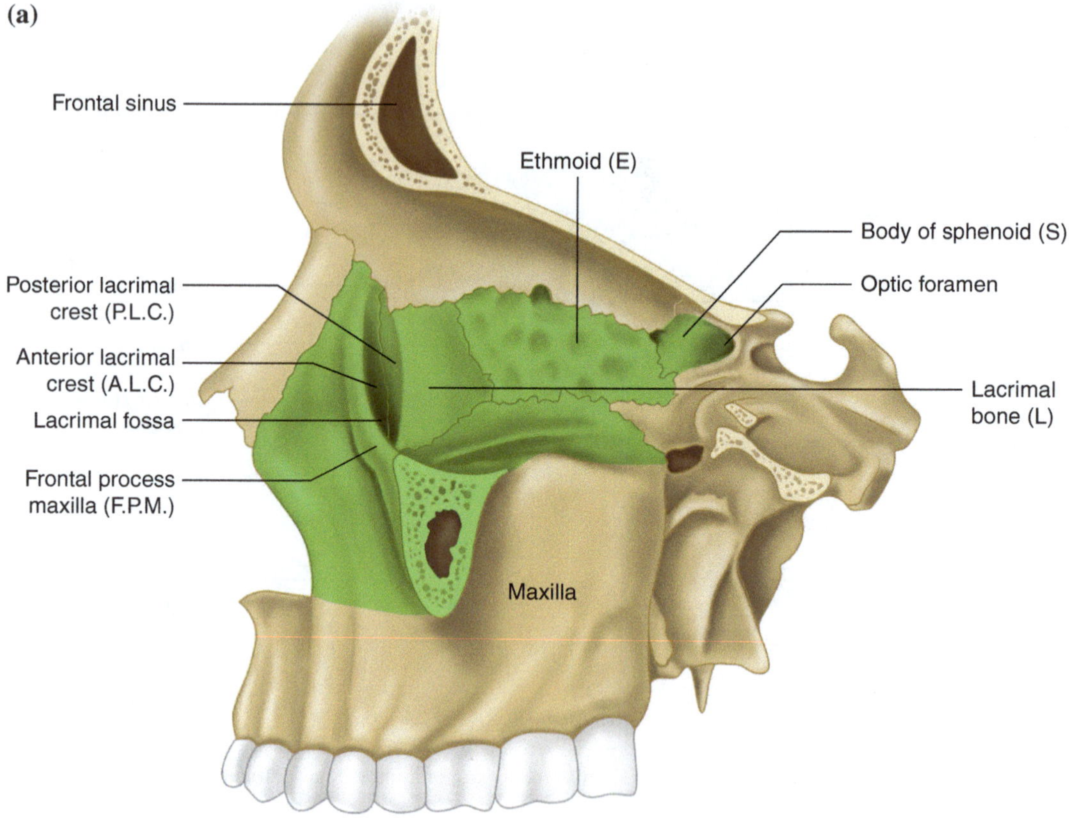
Frontal sinus
Ethmoid (E)
Body of sphenoid (S)
Posterior lacrimal crest (P.L.C.)
Optic foramen
Anterior lacrimal crest (A.L.C.)
Lacrimal bone (L)
Lacrimal fossa
Frontal process maxilla (F.P.M.)
Maxilla

(b)

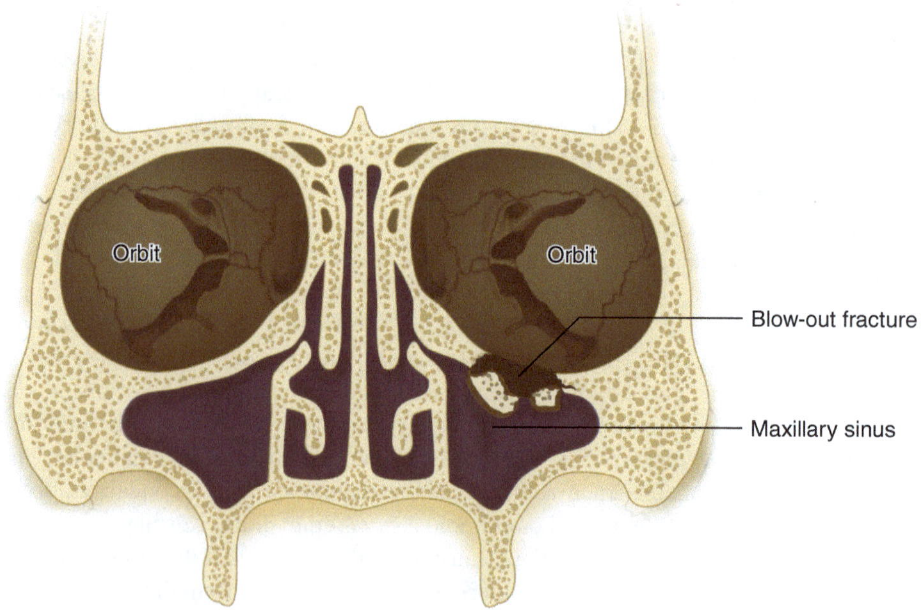
Orbit
Orbit
Blow-out fracture
Maxillary sinus

◀ **Fig. 1.8** **a** Medial wall orbit (thinnest) four bones: (1) frontal process of maxilla; (2) lacrimal; (3) ethmoid (E); (4) body of sphenoid(s). **b** Coronal computed tomographic scan of left blow-out fracture. **c** Mechanism of blow-out fracture; sudden gross increase in intraorbital pressure by a squash ball/or clenched fist may cause ecchymosis of the lids, restricted elevation of the eye, binocular diplopia, anesthesia of the upper lip, hypotropia, or the nose may bleed after blowing if the medial wall is also ruptured; treatment and referral to an ophthalmologist are urgent

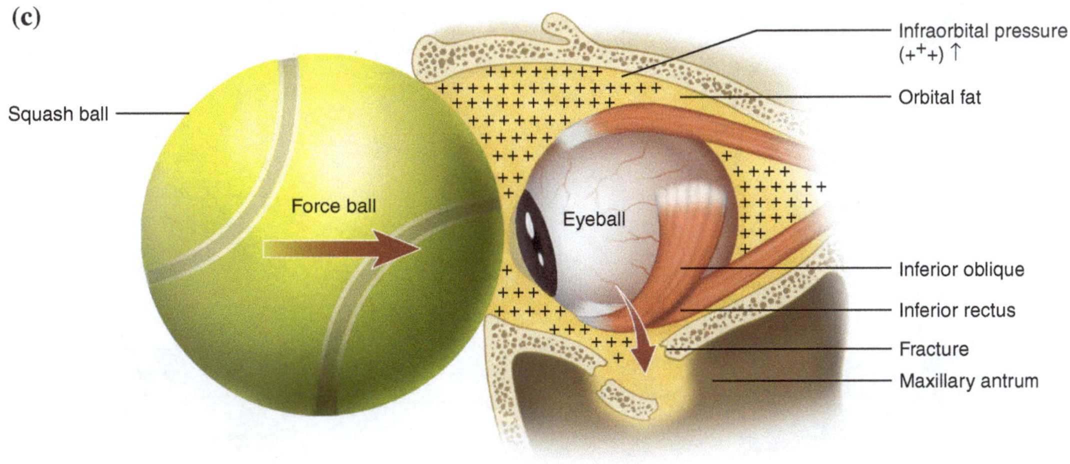

Fig. 1.8 (continued)

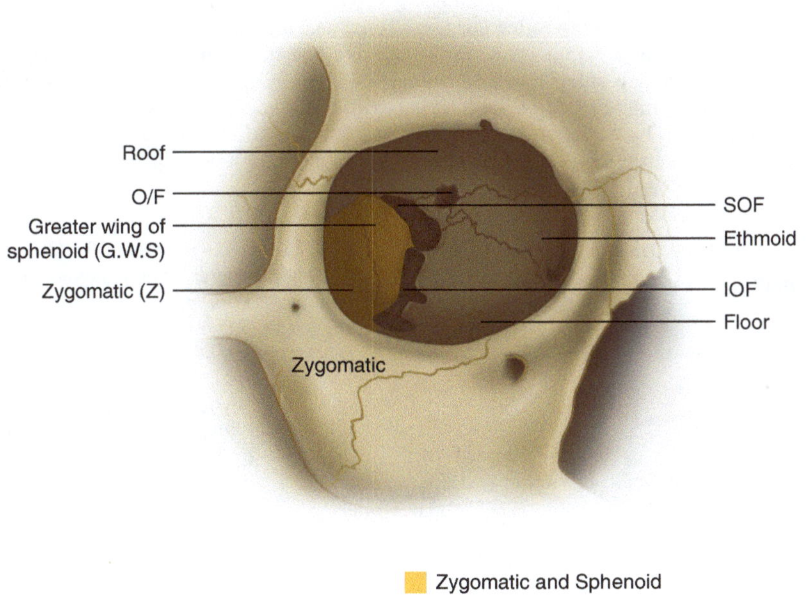

Fig. 1.9 Orbit of lateral wall (two bones)

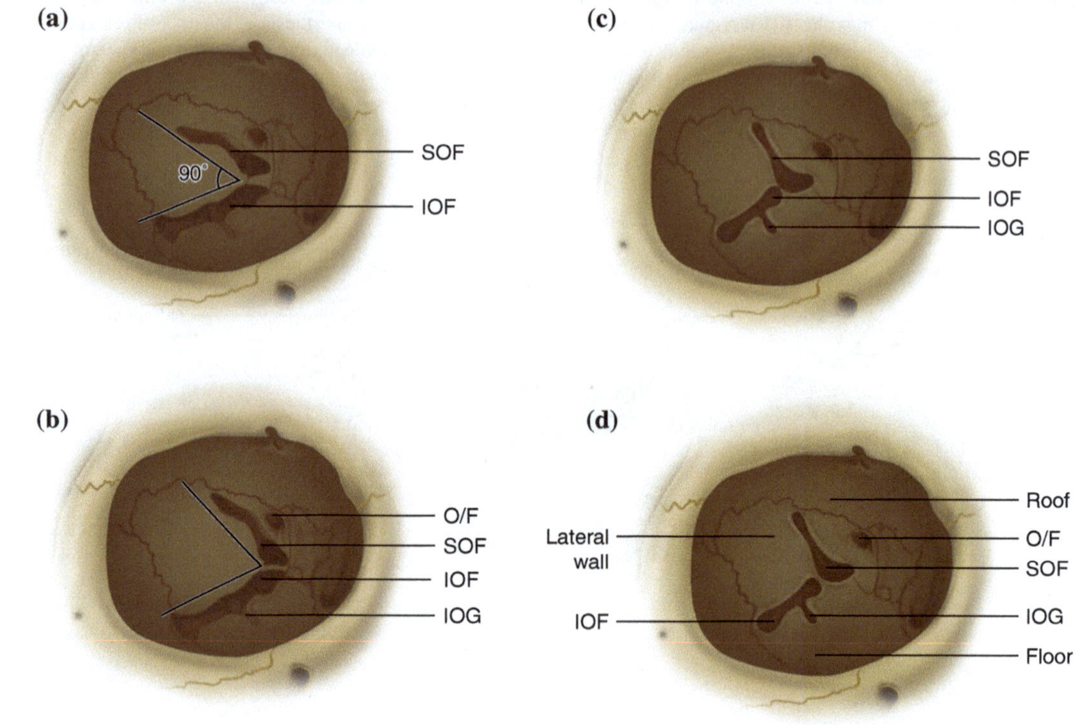

Fig. 1.10 **a–d** Let us have some fun in anatomy. If you study Fig. 1.4, you may notice that the angle between the SOF and IOF is roughly 90°, and the optic foramen lies almost at the level of the IOF and internal to the junction of the SOF and IOF. If we draw a mirror image of the alphabet letter "L," the vertical limb of L can be imagined to be the SOF and the oblique limb of L to be the IOF; a small line may be imagined to be the infraorbital groove (IOG), and the circle optic foramen (OF) can be imagined to be the optic foramen (**a** and **b**). **c** If you look again at Fig. 1.4 and copy the real contours of the SOF along the vertical line, the contours of the IOF along the horizontal line, and the contours of IOG and OF, you will get a figure resembling **d** that may be imagined to represent SOF, IOF, IOG, and the OF. The lateral wall of the orbit will be the roof of the orbit and below that will be the floor. It can help to have a mental picture of the roof, the floor, and the lateral wall

headaches in susceptible patients. Let us have some fun in anatomy (Fig. 1.10 a–d).

Suggested Reading

Parsons' Diseases of the Eye, edn 16. Revised by Miller SJH. London: Churchill Livingstone; 1979. pp. 1–14; 545–60

Riordon-Eva P, Cunningham E (2011) Vaughan and Asbury's general ophthalmology, edn 18. McGraw Hill, New York, pp 1–26

Trevor-Roper PD (1974) The eye and its disorders. Blackwell Scientific Publications, London. pp 26–44, 323–34

Wolff's E (1976) Anatomy of the eye and orbit, edn 7. Revised by Warwick R. London: HK Lewis; 1976; pp. 1–29; 30–180D

2 The Eyeball: Some Basic Concepts

Keywords
Concentrically arranged layers · Cornea · Sclera · Uveal tract · Iris · Ciliary body · Choroid · Posterior scleral foramen · Lamina cribrosa · Anterior chamber · Posterior chamber · Vitreous chamber · Ciliary processes · Secretion · Ultrafiltration · Circulation of aqueous · Intraocular pressure · Microscopic outlet channels of aqueous · Inflow of aqueous · Outflow of aqueous

The eyeball consists of three concentrically arranged layers. The frontal one-sixth of the outer wall is formed by a transparent dome called the cornea. The back five-sixths of the outer wall of the eye is formed by a fibrous layer protective in function called the sclera. The sclera is quite thin at its front where the rectus muscles are inserted, and quite thick at its back, which explains why traumatic rupture of the eyeball is common there. It is richly supplied by sensory nerves (ophthalmic division of the trigeminal nerve); therefore its inflammation, or scleritis, is quite painful. The outer surface of the anterior sclera is covered by a thin layer of elastic tissue, called the episclera; this contains numerous blood vessels that nourish it. The brown pigment layer on its inner surface is called the lamina fusca, which forms the outer layer of the suprachoroidal space. At its back, the sclera has a large opening called the posterior scleral foramen, bridged by the lamina cribrosa; the latter has microscopic holes, through which axons of ganglion cells of the retina leave the eyeball as the optic nerve. Around the optic nerve the sclera is penetrated by long and short posterior ciliary arteries and long and short ciliary nerves. The four vertex veins which drain the choroid exit through the sclera.

The middle coat of the eyeball is called the uveal tract. It consists of the iris on the front, the ciliary body in the middle, and the choroid at its back. The iris is like a circular diaphragm placed in a coronal plane in front of the lens with a central circular hole called the pupil. In addition to controlling the amount of light entering the eye, the pupil also acts as a communication between the anterior and posterior chambers for circulation of the aqueous. The iris has two unstriated involuntary muscles: (1) the sphincter pupillae at the pupillary margin that constricts the pupil, and (2) the dilator pupillae, which dilates the pupil in an emergency. Because the iris is rich in sensory supply, iritis, inflammation of the iris, is painful. The periphery of the iris is attached to the middle part of the ciliary body by a thin root that gets torn in severe contusions of eye, deforming the normal circular pupil.

M.W. Ansari and A. Nadeem, *Atlas of Ocular Anatomy*, DOI 10.1007/978-3-319-42781-2_2

The space between the back of the cornea and the front of the iris is the anterior chamber (AC). The space between the back of the iris and the front of the suspensory ligament is called the posterior chamber (PC). The space between the back of the lens and the front of the retina is called the vitreous chamber (VC). The AC and PC normally both have a transparent liquid called the aqueous humor, which gives nutrition to the cornea and lens. The vitreous chamber has an avascular gelatinous body called the vitreous, which fills the space between the lens, the retina, and the optic disc.

The ciliary body is triangular in shape and located in the horizontal section of the eye; the base of triangle is in front and the apex is at the back. Its anterior 2 mm has 70 fingerlike processes in a circular fashion with a zonule on the medial side. This part of the ciliary body is called the pars plicata, which is continuous behind with a 6 mm long part with a plane surface called the pars plana. The ciliary processes and pars plana of the ciliary body are lined by a two-layered epithelium. The epithelium covering the ciliary processes is the site of an active process called "secretion" in which the aqueous is mainly secreted in the PC. From there it goes to the AC via the pupil, and then to the periphery of the AC, called the angle of the anterior chamber, which has microscopic outlet channels of aqueous called the trabecular meshwork, the canal of Schlemm, 30 collecting trunks, and 12 aqueous veins. These channels ultimately drain the aqueous to the episcleral venous plexus. This circulation of the aqueous creates a pressure inside the eyeball called intraocular pressure (IOP), which maintains the shape of the eyeball essential for refraction at the cornea. A balance between inflow and outflow of aqueous normally maintains the IOP between 12 and 20 mm Hg. For maintenance of a normal IOP, the angle of the AC must have an optimum width so that the aqueous has adequate access to its microscopic outlet channels. The width of the angle of the AC depends upon how deep the AC is, and its depth depends upon the contour of the iris. A convex forward iris will create a narrow angle as in the swelling of a hypermature cataract, or anterior traumatic dislocation of the lens. In addition, a small hypermetropic eye may have a shallow AC.

2.1 Basic Concepts

The eyeball resembles a ping-pong (table tennis) ball with a transparent dome in front called the cornea. When a patient looks at a distant object with both eyes open (the so-called primary position of the eyeball), you can see the cornea as a shiny dome in the center surrounded by the white of the eye on its two sides, called the sclera, which is normally covered by a moist mucus membrane called the conjunctiva because it connects the sclera to the inside of the lids. This part of conjunctiva is called the bulbar conjunctiva because it covers the front of the eyeball. If you trace this bulbar conjunctiva superiorly and inferiorly, you will find that it jumps to the inside of upper and lower lids in the form of a horizontal loop called the fornix. There is superior and an inferior fornix nearby, in which the accessory lacrimal glands of Krause and Wolfring are found. The circular periphery of the cornea, which constitutes the anterior one-sixth of the eyeball, is continuous behind with the sclera. The junction is called the limbus, which has a circular outline. The sclera, the outermost layer of the eyeball, is fibrous in structure and protective in function. It constitutes the posterior five-sixths of the eyeball. It has a rich sensory supply and therefore its inflammation, called scleritis, is painful. In the primary position of the eyeball the lower margin of the upper lid just touches the upper limbus (Fig. 2.1). The gaps between the two lid margins, the palpebral fissures, are normally equal on the two sides. In cases of high myopia or thyrotoxicosis, a scleral rim may be seen just below the upper lid margin or above the lower lid margin (Fig. 2.2a). In another situation, one eyeball may be pushed down and out by a mucocele of the frontal or ethmoid sinus, again creating a narrow palpebral fissure in the primary position (Fig. 2.2b). Mechanical pushing of the eyeball from the orbit can occur in space-occupying lesions of the orbit, such as thyroid orbitopathy, glioma of the optic nerve, or a gross traumatic retrobulbar

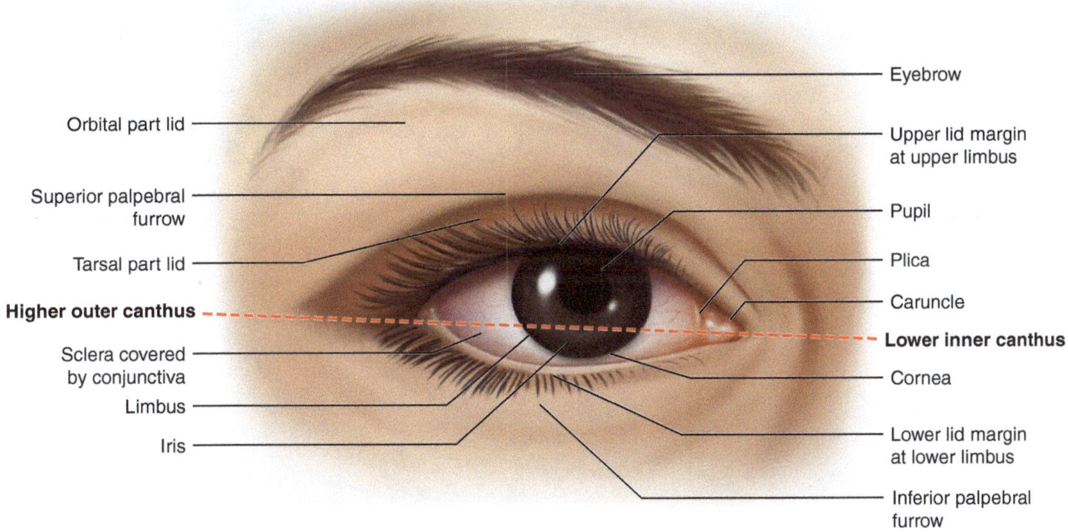

Fig. 2.1 The eye in a primary position. Note the *upper* and *lower* limbi along the lid margins

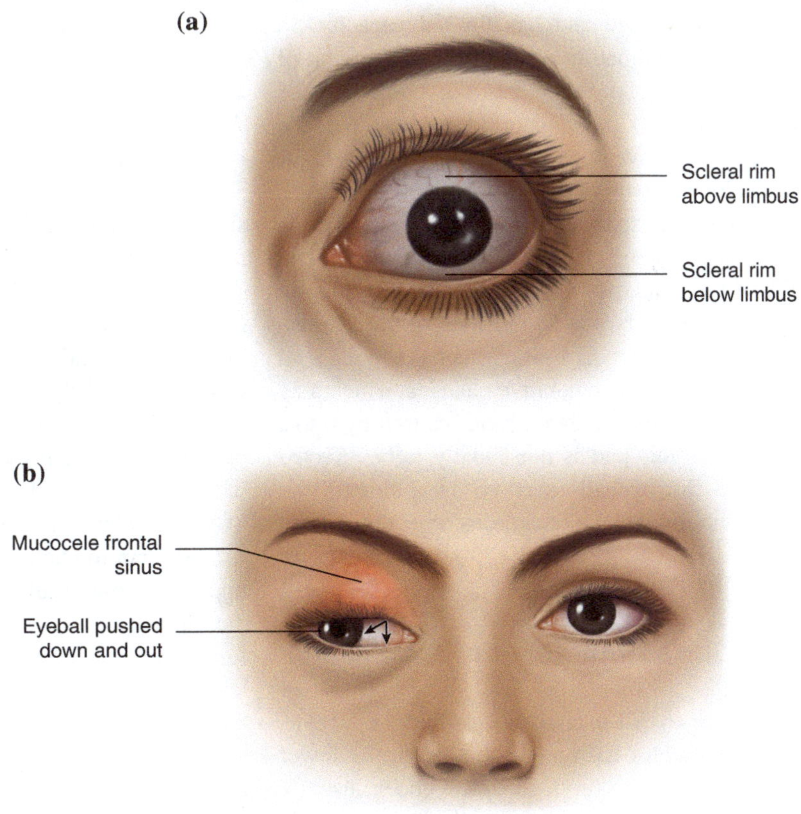

Fig. 2.2 **a** and **b** Mucocele of the right frontal sinus. The right eyeball is pushed down and out-proptosed obliquely

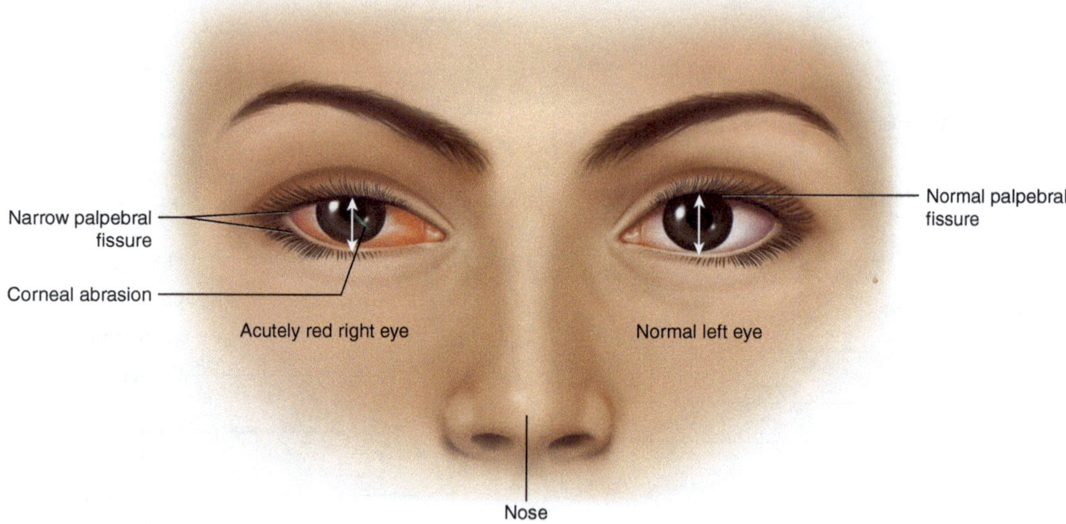

Fig. 2.3 Narrow palpebral fissure in a red eye

hemorrhage behind the eyeball because the orbit is a closed space and has limited capacity. In cases of inflammation of the anterior segment of the eye (e.g., conjunctivitis, keratitis, scleritis, and anterior uveitis), the palpebral fissure on that side becomes narrow because of a reflex spasm of the palpebral part of orbicularis oculi (Fig. 2.3). The most exposed part of the globe is just below the center of the cornea. When we close both lids, the eyeball moves up. Even when the lids are shut, the part below the center of the cornea remains uncovered. That explains why it is a common site of degenerative changes, which may result from exposure. In injuries caused by burns or caustics this part is most likely to be affected because the eyeball goes up in a threatening reflex motion when a dangerous object is in front of the eye.

2.2 Surface Anatomy of the Eye

The upper eyelid extends above to the eyebrow, which separates it from the forehead; the lower eyelid passes into the skin of the cheek, usually without a line of demarcation. The upper lid is more mobile than the lower. The upper lid has an elevator muscle called the levator palpebrae superioris. The lid margins meet laterally to form the outer canthus and medially to form the inner canthus. When the eyes are open widely the outer canthus measures about 60° and is about 5–7 mm from the orbital margin. It is at a slightly higher level than the medial canthus.

The inner canthus encloses a small area of a skinlike structure called a caruncle. Touching the lateral edge of the caruncle is a semilunar fold of conjunctiva called the plica semilunaris. The caruncle is normally covered by a microscopic layer of tears called the lacrimal lake. The lateral canthus is directly placed on the globe. The portions of the eye normally visible in the palpebral opening are the shiny cornea in the center, behind which is the colored iris surrounded by a triangle of white sclera on the lateral side and by crescentie sclera medially. When the eye is open, the upper lid shows a sulcus along the upper border of the upper tarsus called the superior palpebral furrow, produced by the pull of the tendon of the levator palpebrae superioris. It disappears when the lids are closed. The corresponding furrow in the lower lid is illmarked. The eyeball is made of three concentric layers; the outermost layers are the cornea in the front and the sclera in the back. The transparent dome of the cornea forms the front one-sixth of the outer wall, and the posterior

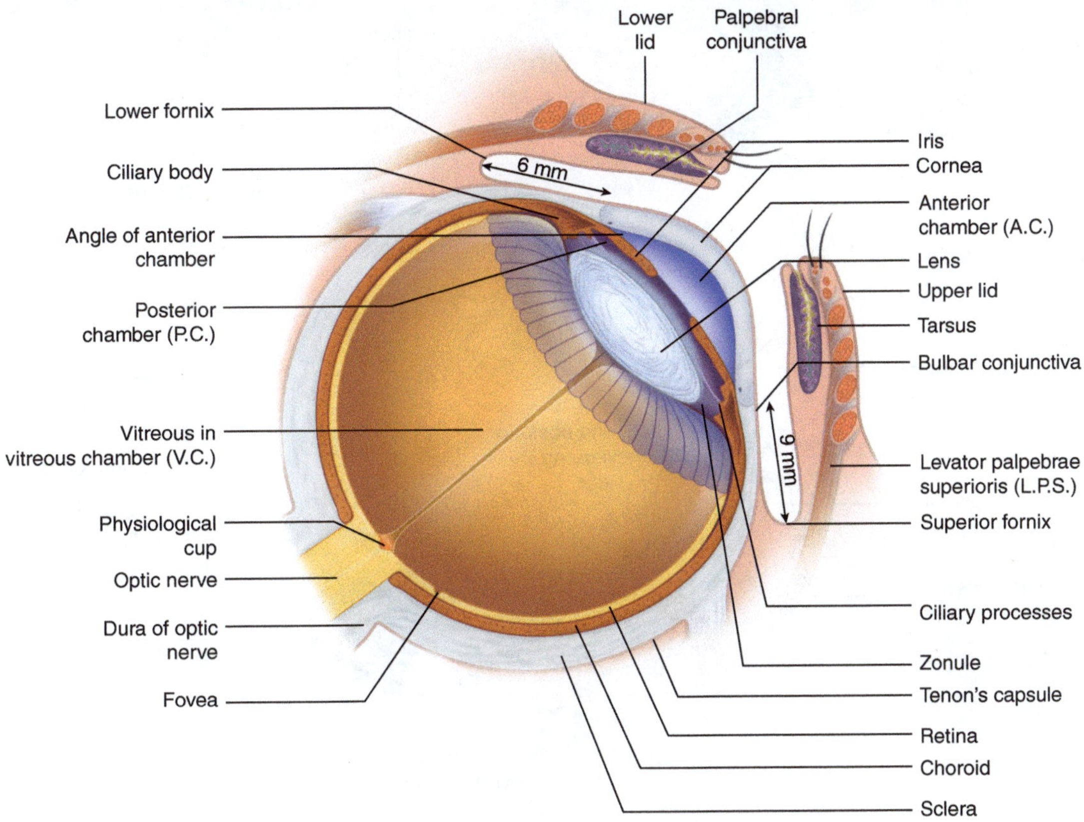

Fig. 2.4 Horizontal section of the eyeball. Note that the superior fornix is higher than the inferior fornix. Note also the three chambers: the *AC*, *PC*, and *VC*. There are three layers of the eyeball and the superior and inferior fornices are visible. See the continuity of the sclera with the dura of the optic nerve

five-sixths of the outer wall is formed by the sclera (Fig. 2.4). The middle coat is the uveal tract, which is vascular in structure and nutritive in function. It has three parts: the iris in the front and the ciliary body in the middle, which is continuous with the choroid in the back. The innermost nervous layer is called the retina. The iris and ciliary body are also called the anterior uvea by virtue of their position, and the choroid is often called the posterior uvea. Their description follows, starting with the sclera.

2.3 The Sclera

The fibrous outermost layer of the sclera of the eyeball is thinner in front at the insertions of the rectus muscles and thicker at the back. That explains why rupture of the eyeball occurs mostly in that area in severe contusion injuries. At the back of the sclera is a large gap called the posterior scleral foramen; this is bridged by layers of connective tissue called lamina cribrosa containing microscopic holes through which axons of the ganglion cells of the retina leave the eyeball as the optic nerve. Figure 2.5 gives all the details of the scleral shell, which has to be pierced by arteries, veins, and nerves supplying the eyeball. The two superior and two inferior vertex veins pierce it 4 mm behind the equator of the eye. The long and short posterior ciliary arteries and nerves have to pierce it to supply the eyeball. From the limbus to the fornix, the sclera is covered by bulbar conjunctiva, which is kept moist by a film of tears. The outer surface of the limbus to the optic nerve is covered by a part of

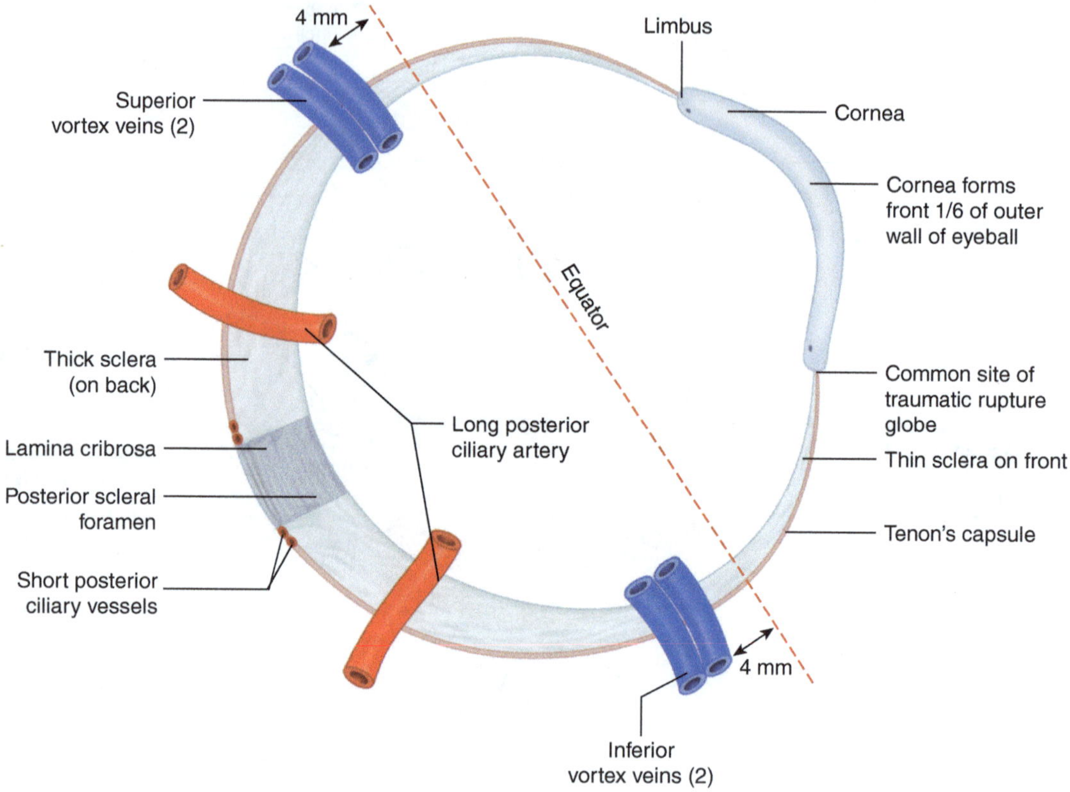

Fig. 2.5 The scleral shell. Note the blood vessels and nerves piercing it. The posterior five-sixths of the outer layer of the eyeball is formed by the sclera, which is very thin at insertions of the rectus muscles; therefore traumatic rupture mostly occurs there. The optic nerve passes through its back hole, which is called the posterior scleral foramen. Note the Tenon capsule covering it. Because it is rich in sensory supply, scleritis is very painful

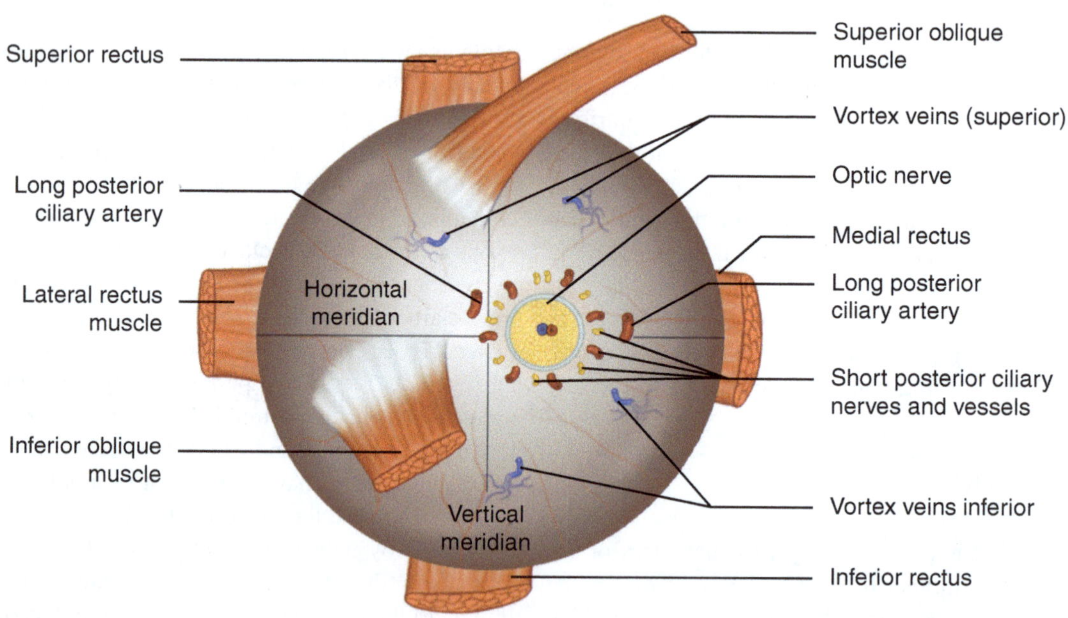

Fig. 2.6 The left eyeball from behind

the orbital fascia called the Tenon capsule, which is continuous at its back with the dura of the optic nerve. The sclera is richly supplied by sensory nerves; therefore its inflammation, called scleritis, is painful. The four rectus muscles—the superior rectus (SR), the inferior rectus (IR), the medial rectus (MR), and the lateral rectus (LR)—are inserted on the front of the sclera at variable distances from the limbus in front of the equator of the eye. The superior oblique (SO) and inferior oblique (IO) muscles are inserted at the back behind the equator of the eye (Fig. 2.6).

2.4 Uveal Tract: Iris, Ciliary Body, Choroid (Figs. 2.7 and 2.8)

The iris is a circular diaphragm placed in a coronal plane in front of the lens forming the posterior boundary of the space between the back of the cornea and the front of the iris called the anterior chamber (AC). It has a central circular hole called the pupil, which controls the amount of light entering the eye and also acts as a communication between the anterior and posterior chambers for the aqueous. It has a peripheral ciliary border and a central pupillary border. The ciliar border houses the major arterial circle of the iris formed by the joining of two long posterior ciliary arteries, which send radial branches to supply the iris and also to the fenestrated capillaries of the ciliary processes. The attachment of the periphery of the iris to the ciliary body is called the root of iris, which is thinner than the rest of the iris. This explains why a traumatic tear of the iris may occur at its root, which creates a deformed pupil. The potential space between the back of iris and the front of the zonule is called the posterior chamber (PC). The anterior and posterior chambers both have a fluid called the aqueous humor, which is secreted chiefly in the posterior chamber by fenestrated capillaries of the ciliary processes. The iris is lined by two-layered epithelia that are a continuation of the two-layered epithelia lining the ciliary body.

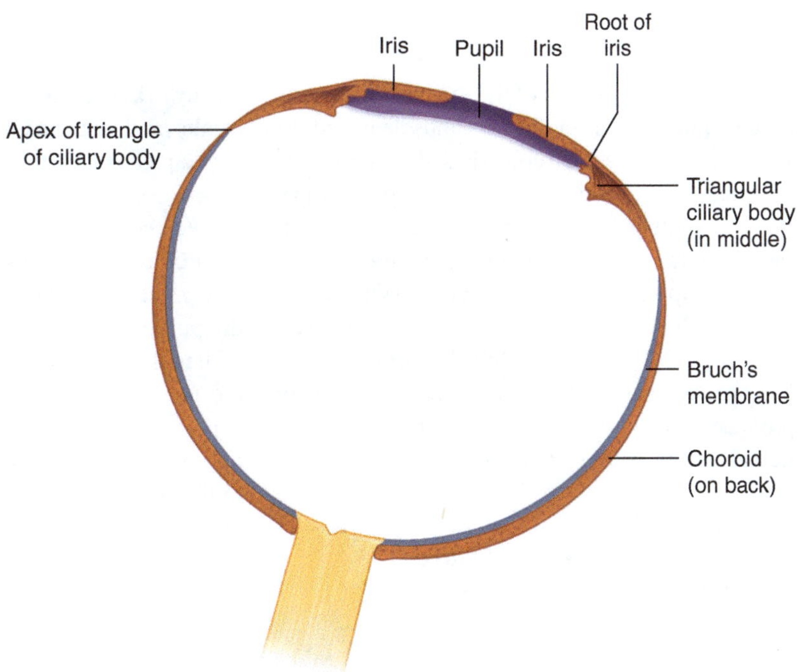

Fig. 2.7 The middle coat of the eyeball: (1) the uvea; (2) the iris (*front*); (3) the ciliary body (*middle*); (4) the choroid (*back*). It is vascular in structure and nutritive in function. The suprachoroidal space is between the choroid and the sclera. It is partly used for aqueous drainage

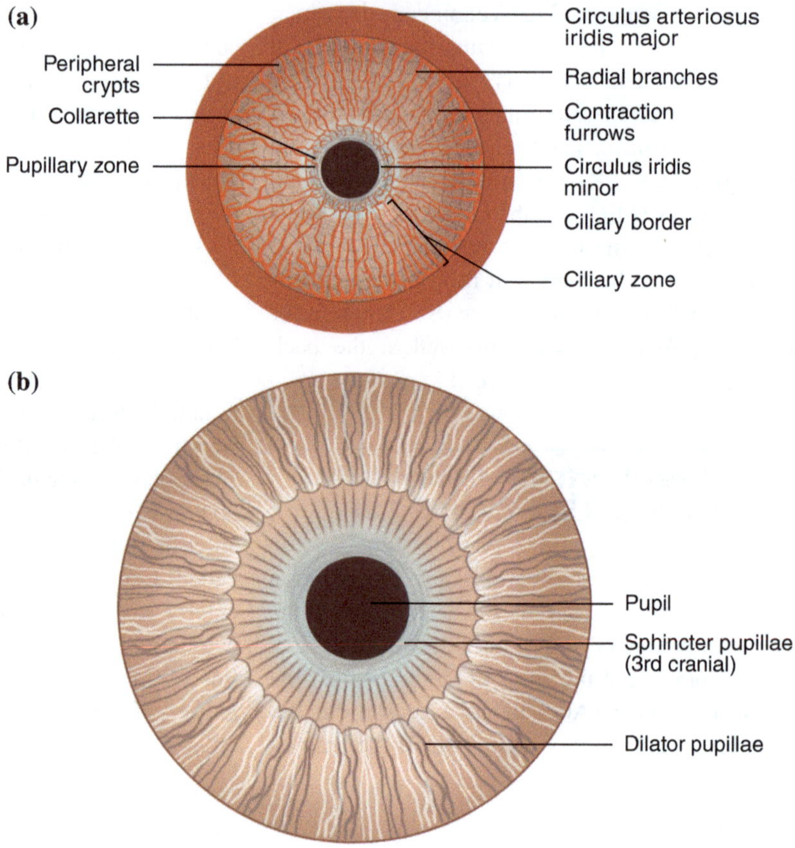

Fig. 2.8 **a** Iris blood vessels (anterior surface). **b** Involuntary muscles of the iris

The pupillary margin of the iris has a circularly arranged unstriated involuntary muscle called the sphincter pupillae that constricts the pupil. It is supplied by a third cranial nerve (parasympathetic). Another involuntary muscle in the iris is radially arranged. It is called the dilator pupillae and is supplied by the cervical sympathetic nerves, which dilate the pupil in emergencies. Pupillary size is principally determined by a balance between the constrictor and dilator pupillae. The iris is richly supplied by the sensory nerves and therefore iritis is painful.

2.5 Anatomy of the Angle of the Anterior Chamber

The periphery of the AC is called its angle. It lies at the junction of the peripheral cornea and the root of the iris. In it are situated the microscopic outlet channels of the aqueous, i.e., the trabecular meshwork, the canal of Schlemm, and collecting trunks and aqueous veins that ultimately drain the aqueous to the episcleral venous plexus (Fig. 2.9).

We know that the aqueous is chiefly secreted in the PC from the fenestrated capillaries of the ciliary processes. From the PC it goes to the AC via the pupil; it then reaches the periphery of the AC (called the angle of the anterior chamber) to reach the microscopic outlet channels of the aqueous. It is clear that the angle of the AC must have an optimum size for adequate access of the aqueous to the outlet channels there. Naturally, the width of the angle of the AC depends on the depth of the AC. A shallow AC has a narrow angle. The size of the angle cannot be directly assessed on slit-lamp examination because of total internal reflection. Therefore, it can be assessed only using an instrument called a gonioscope, which removes total internal reflection. This is done under surface anesthesia with a gonioscope by an

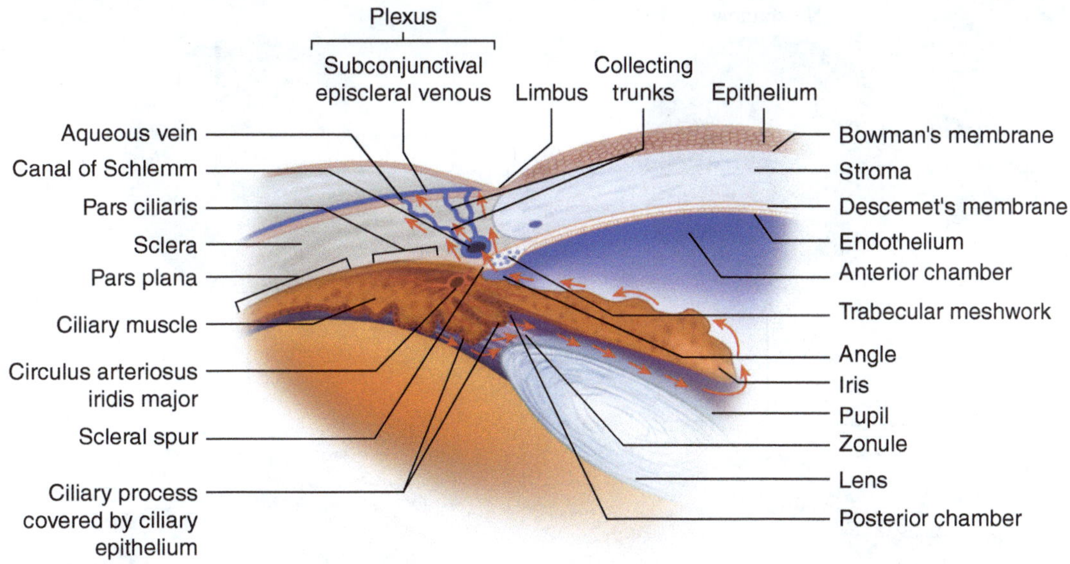

Fig. 2.9 The microscopic outlet channels: (1) trabecular meshwork; (2) canal of Schlemm; (3) aqueous veins (there are 12); (4) collecting trunks (there are 30)

ophthalmologist on a slit lamp. Any condition that makes the AC shallow, such as a small hypermetropic eye, anterior dislocation of the lens by injury, or a hypermature swollen senile cataract may cause acute congestive glaucoma (ACG), in which intraocular pressure suddenly rises to almost 60 mm Hg or above, causing stretching of the corneal nerves, severe pain, and vomiting; this is an ophthalmic emergency. Dilatation of the pupil in an eye of a shallow AC also can cause crowding of the iris tissue in the angle and Acute Congestive Glaucoma. Therefore before dilating the pupil in any eye, it must be checked that the AC is not shallow. This can be clinically assessed by throwing torchlight from the lateral side into the eye. If the iris is convex at the front (shallow AC), you will see a crescentic shadow called the iris shadow at the pupillary margin. This is the iris shadow test (Fig. 2.10).

2.6 Ciliary Body

The middle part of the uveal tract is continuous in front with the iris and at the back with the choroid. It is triangular in cross section. Its front part (about 2 mm) has about 70 fingerlike processes called ciliary processes that have fenestrated capillaries from the circulus iridis major in the root of the iris. It has a zonule (the suspensory ligament of lens) on its medial side. These ciliary processes are in a circle and are attached to the zonular fibers around the lens. To have a three-dimensional concept, imagine yourself standing inside the vitreous behind the lens looking in front, where you will see five concentric circles (Fig. 2.11). Innermost is the edge of the lens; next is the zonule, connecting the lens to the zonule; third is the pars plicata, part of the ciliary body with ciliary processes; fourth is the pars plana, the third part of the ciliary body that is plain; and fifth is the margin ora serrata, where the retina ends.

The ciliary body has an unstriated involuntary muscle called the ciliary muscle with a combination of circular, radial, and longitudinal fibers. The circular fibers on contraction relax the zonule of the lens, which in turn relaxes the capsule of the lens. This produces an increased curvature of the anterior pole of the lens because of moldable lens matter in people below 40 years of age in whom the lens matter is soft. This increased curvature of the anterior pole of the

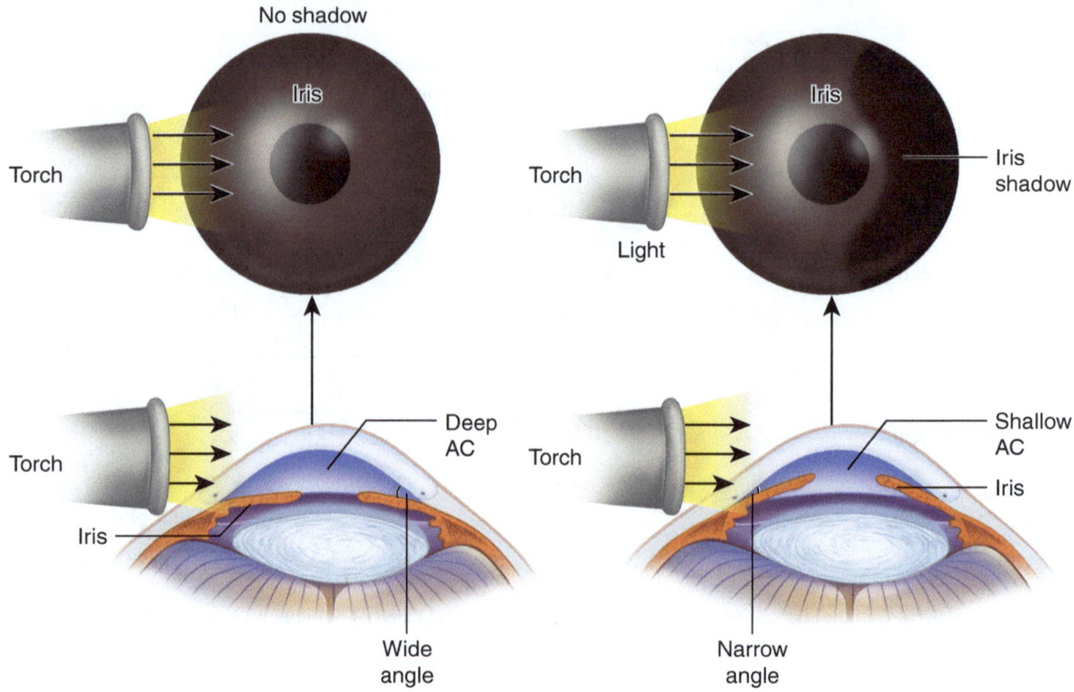

Fig. 2.10 Iris shadow test for narrow angle of the anterior chamber

Fig. 2.11 Imaginary posterior view of the ciliary body, zonule, lens, and ora serrata while sitting in vitreous

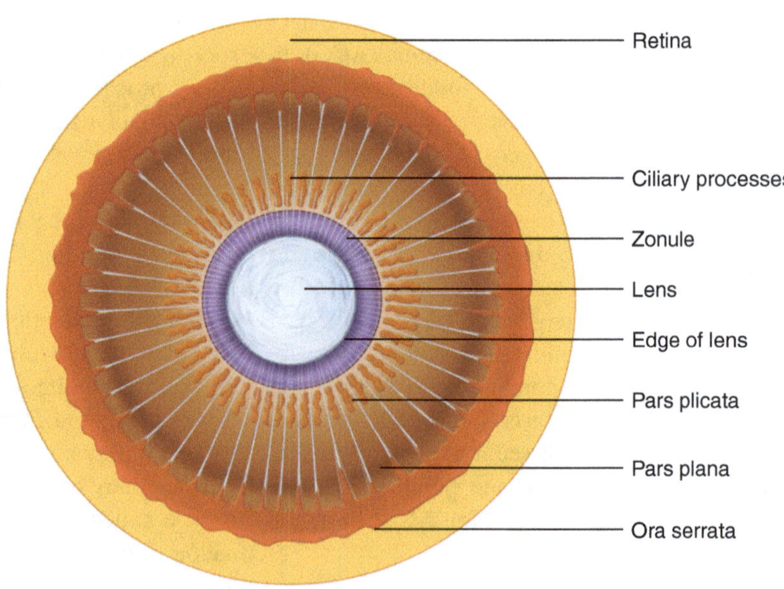

lens helps to focus on a near object from which the incoming rays of light are divergent and need to be converged by increased curvature of the anterior pole of lens. This device in young people is called accommodation, which becomes weak in old people; if they want to read small letters they have to use reading glasses because their accommodation is poor.

The longitudinal fibers of ciliary muscle are inserted into the trabecular meshwork. They

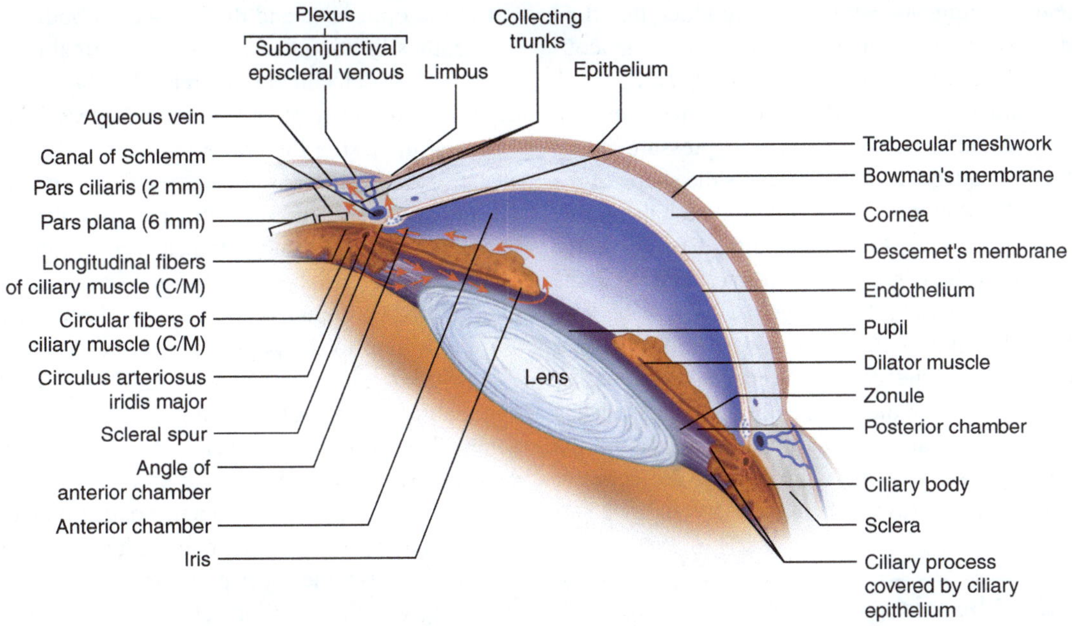

Fig. 2.12 Ciliary body, ciliary processes, and circulation of the aqueous (*red arrows*)

dilate the trabecular spaces to improve the aqueous outflow, thus keeping intraocular pressure within normal limits. The inward extension of the sclera between the ciliary body and the Schlemm canal is called the scleral spur, to which the iris and the ciliary body are attached (Fig. 2.12). The ciliary body is richly supplied by sensory nerves; therefore cyclitis, its inflammation, is quite painful.

2.7 Circulation of the Aqueous and the Shape of the Eyeball

We have already stated that a sterile fluid called the aqueous humor is chiefly secreted by fenestrated capillaries of the ciliary processes into the PC. These ciliary processes are covered on their inner surfaces by two layers of epithelium called the ciliary epithelium. Two processes are at work in the formation of the aqueous: one is an ultrafiltration of plasma in the stroma of the ciliary processes (20 %); the other is an active process of secretion (80 %) conducted by the metabolic activity of the cells of the ciliary epithelium and resulting in secretion of sodium and ascorbic acid. Secretion is an active process that needs energy. Aqueous is a transparent clear liquid that fills the ACs and PCs of the eye. Its volume is about 250 μL and its rate of production, subject to diurnal variation, is about 2.5 μL/min. Its osmotic pressure is slightly higher than that of plasma. The composition of the aqueous is similar to that of plasma except for much higher concentrations of ascorbate, pyruvate, and lactate and lower concentrations of protein, urea, and glucose.

Entering the posterior chamber, the aqueous passes through the pupil into the AC and then to the trabecular meshwork in the angle. During this period, there is some differential exchange of components of blood in the iris.

The trabecular meshwork is composed of collagen and elastic tissue that form a filter with a decreasing pore size as the canal of Shlemm is reached. Trabecular spaces in the trabecular meshwork open into the canal of Schlemm. Passage of the aqueous into the canal of Schlemm depends upon cyclic formation of transcellular channels in the endothelial lining of the cornea. There are efferent channels from the canal of Schlemm (about 30 collector channels and 12 aqueous veins) that conduct the fluid directly into the venous system. In brief, aqueous comes from blood veins (INFLOW) and goes to the blood (OUTFLOW). A balance between inflow and outflow maintains a pressure inside the eye called IOP that maintains the shape of the eyeball. Otherwise, the eye would be indented by atmospheric pressure. This support of the shape of the eyeball is necessary for refraction at the cornea. The walls of the eyeball have to be kept tight by this intraocular pressure for proper refraction at the cornea. Some aqueous passes (20 %) between the bundles of the ciliary muscle into the suprachoroidal space and then to the venous system of the ciliary body, the choroid, and the sclera (uveoscleral flow).

2.8 Choroid

Between the retina and the sclera is the choroid, the posterior segment of the uveal tract, composed of three layers of blood vessels—large, medium, and small. The innermost layer is known as choriocapillaries, which supply nutrition to the outer third of the retina by diffusion. Between the sclera and the choroid is the suprachoroidal space, through which 20 % of the aqueous is drained. The rest of the 80 % of the aqueous is drained to the episcleral venous plexus. Posteriorly, the choroid is attached to the margins of the optic nerve, where it ends. On the inner side the choroid is covered by a membrane called the membrane of Bruch. The choroid is supplied with sensory nerve fibers from the trigeminal as well as the autonomic nerves.

2.9 The Retina

The retina is a thin, semitransparent multilayered sheet of neural tissue that lines the choroid. It extends from the optic disc posteriorly up to the end of the ciliary body anteriorly. The anterior end has a ragged edge called the ora serrata. The outer surface is opposed to the retinal pigment epithelium. There is a subretinal space between the pigment epithelium and the sensory retina, but at the optic disc and the ora serrata both are united, thus limiting the spread of subretinal fluid in a retinal detachment. A retinal detachment occurs at the level of the subretinal space. The retina is thinnest at the ora serrata anteriorly and at the fovea on the posterior pole. This explains the common site of traumatic retinal holes in retinal detachment. Embryologically, the retina is continued forward beyond the ora serrata as a two-layered epithelium that lines the ciliary body and the iris up to the pupil.

Most externally, in contact with the pigment epithelium is a neural epithelium with rods and cones, the end organs of vision. Following this, moving inward, are the outer nuclear layer (nuclei of rods and cones), the outer plexiform layer (having synapses), the inner nuclear layer (nuclei of bipolar cells), the inner plexiform layer (having synapses), the ganglion cell layer, and finally innermost the nerve fiber layer composed of axons of ganglion cells coming out of the lamina cribrosa as the optic nerve. All these are bound together by neuroglia formed by special vertical cells called the fibers of Müller. All this is completed by two limiting membranes, the outer perorated by rods and cones and the inner separating the retina from the vitreous. At the posterior pole of the eye, about 3 mm on the temporal side to the optic disc, a specially differentiated spot is called the fovea centralis, which is the avascular and most sensitive part of the retina containing only cones and no other cells. It has a maximum visual acuity of 20/20 because of the one-to-one relationship of its cones to the ganglion cells. There are no rods here. Cones are responsible for color vision; they have three types of pigments for red, blue, and green. The fovea centralis receives its nutrition from neighboring choriocapillaries in the choroid, which is seen as a red spot in the occlusion of the central retinal artery by an embolus. The fovea is surrounded by a small area called the macula leutea or yellow spot, which is more sensitive than other parts of the retina. It is here that the nuclear layers become gradually thinned while ganglion cells are heaped up into several layers. The structural and functional development of the macula is completed by the age of 3 years to reach 20/20 acuity, for which its stimulation by a simultaneous, almost similar bifoveal image is a necessity for proper binocular vision.

The spot where the axons of the ganglion cells come out of the eye through the lamina cribrosa as the optic nerve is called the optic disc, and normally it has a small depression in the center called the physiologic cup (Fig. 2.13a) of the disc, through which central retinal vessels can be seen to emerge by a direct ophthalmoscope (Fig. 2.13b). This cupping becomes deep in a glaucomatous disc (Fig. 2.15).

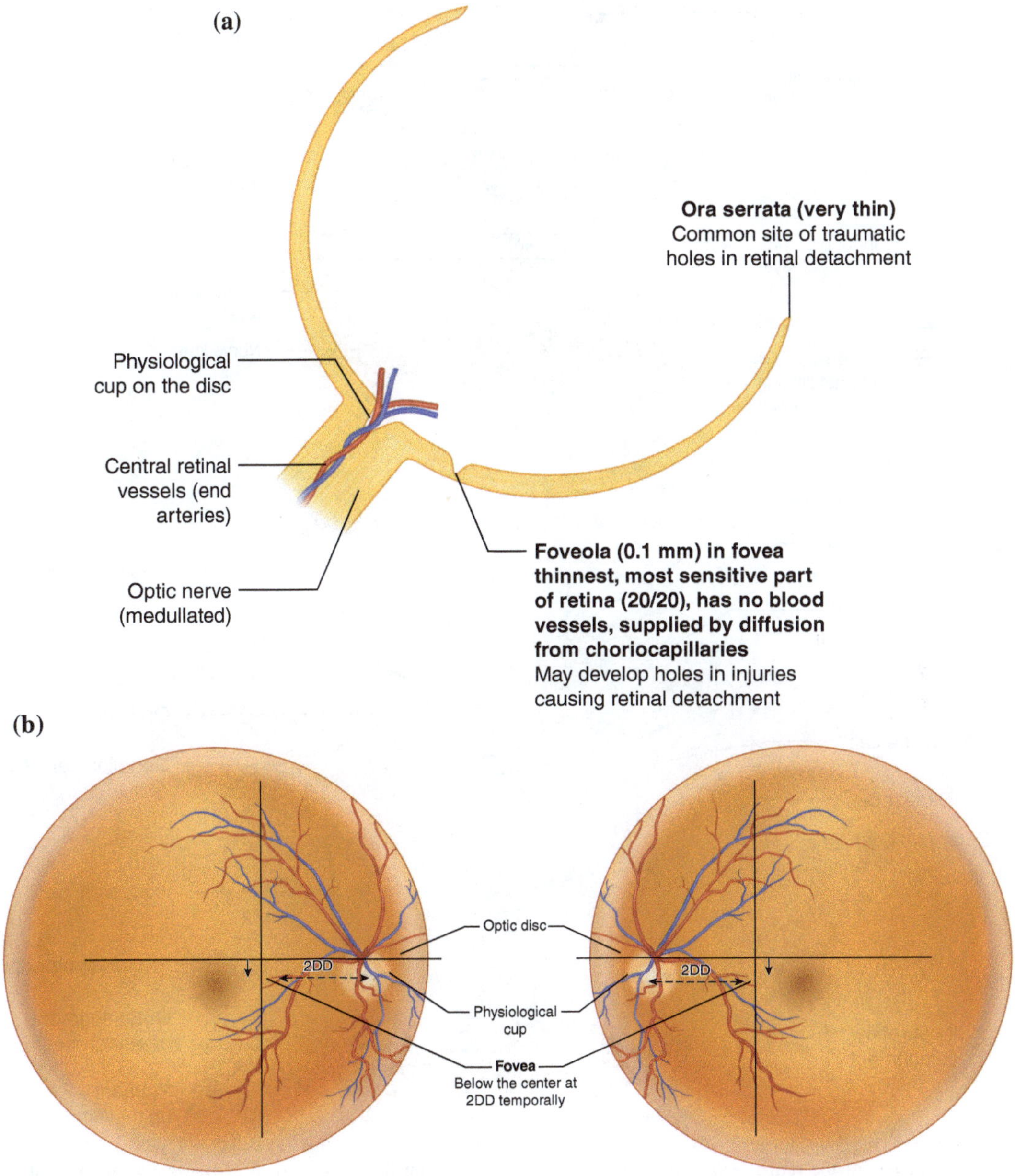

Fig. 2.13 **a** Innermost layer of the eyeball. The retina is thinnest at the ora serrata and the fovea; therefore traumatic holes in the retina mostly occur there. **b** The fundus by direct ophthalmoscope. Note that the fovea is slightly below the center and at two-disc diameter (2DD) away on the temporal side. **c** Layers of the retina

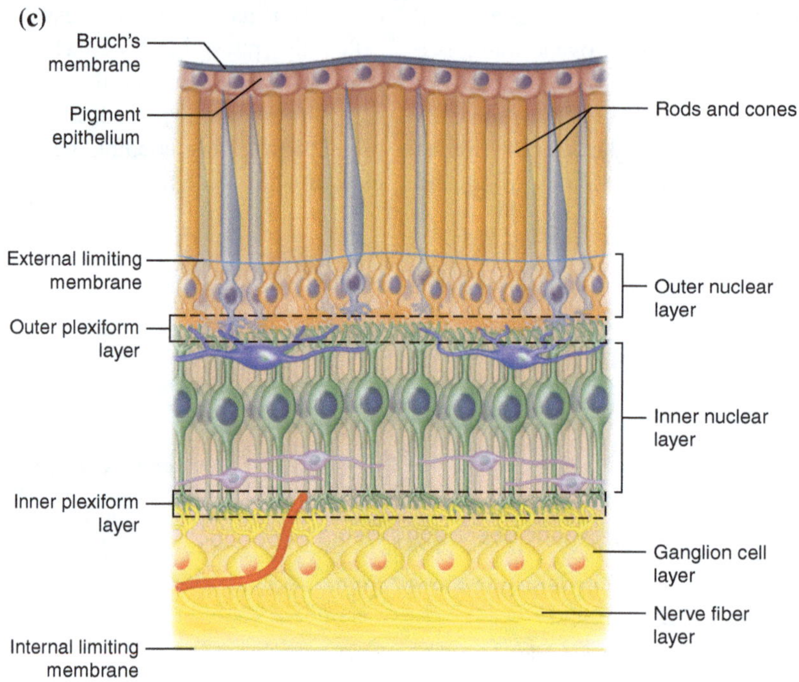

Fig. 2.13 (continued)

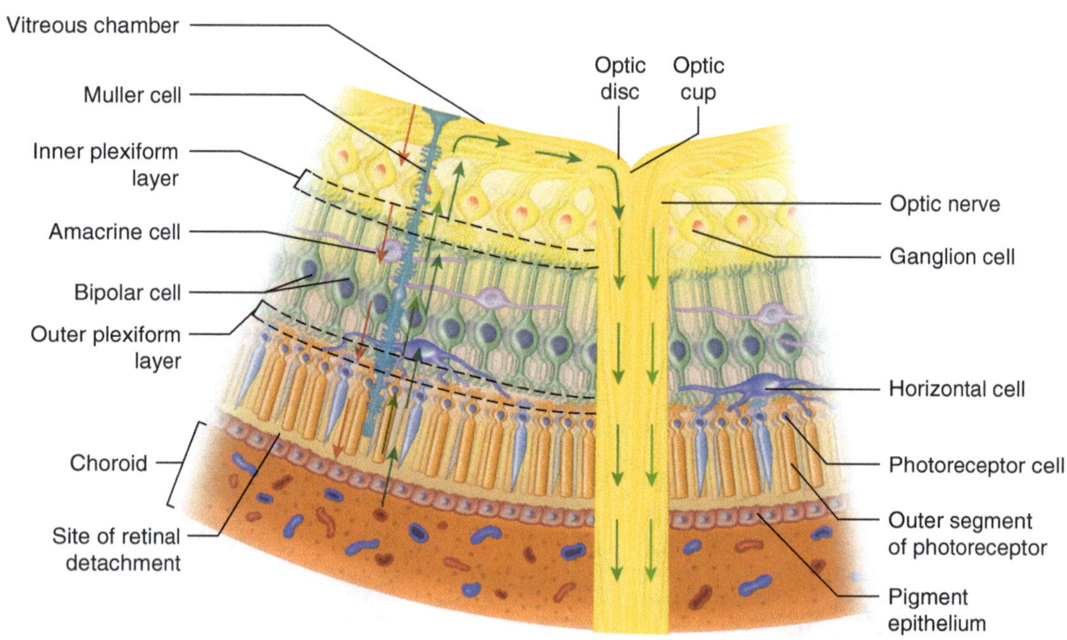

Fig. 2.14 Layers of the retina (*schematic*). *Red arrows* show incident light rays; *green arrows* show the path of the return nerve impulse, which travels back along the visual pathway to the visual center in the occipital lobe

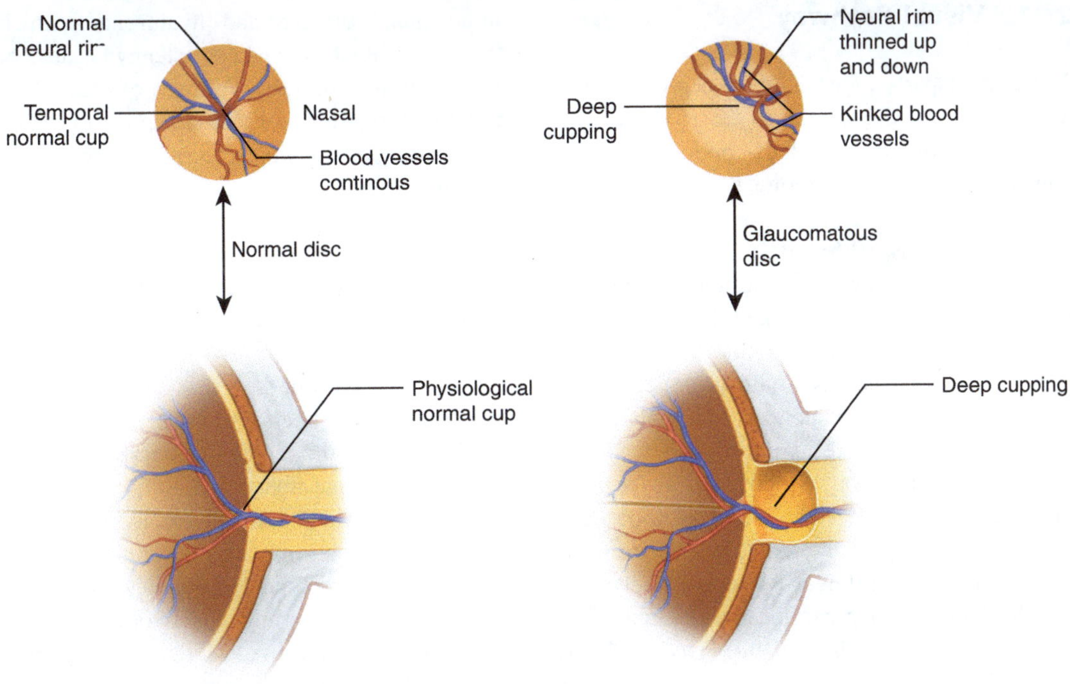

Fig. 2.15 Normal and glaucomatous optic disc

To excite the rods and cones, incident light has to cross the whole retina to reach them. Before striking the photoreceptors, light passes through the entire thickness (*red arrows*) of the retina (Fig. 2.14). The photoreceptor cells initiate a chain of neuronal impulses from the photoreceptor to the bipolar cells and from there to the ganglion cells (*green arrows*, Fig. 2.15), whose axons become the optic nerve when they come out of the eye. This nerve then extends to the brain as a visual pathway.

Embryologically the retina, being a part of the brain, has a dual blood supply. The outer one-third of the retina receives its nutrition from choriocapillaries, and the inner two-thirds of the retina is supplied by the central retinal artery and vein. The central retinal artery is an end artery like other arteries of the brain; therefore its occlusion should result in loss of vision. Because choriocapillaries supply nutrition to the outer one-third of the retina, choroidal disease like choroiditis will affect the retina secondarily.

2.10 The Optic Nerve

We have learned that the optic nerve consists of axons of ganglion cells of the retina that travel via the optic chiasma and the optic tract to the nucleus of the lateral geniculate body, where a synapse occurs and a new neuron takes over the visual nerve impulse to proceed to the visual center of the occipital lobe as optic radiation. The optic radiation fibers end in the visual center, where fusion of two retinal images occurs. That explains why our vision is termed binocular vision. For fusion to occur two retinal images have to be simultaneous, almost similar, and bifoveal (on each fovea).

2.11 Visual Pathway

The cranial parts of each optic nerve join one another above the pituitary fossa to form a four-sided structure called the optic chiasma in which nasal fibers from one side cross to the other side. The optic tract starts from the chiasma, which has 80 % visual fibers and 20 % pupillary fibers. Thus the optic tract of the right side has all the axons of the ganglion cells of the temporal half of the right retina and axons of ganglion cells of the nasal half of the left retina. There axons end in the nucleus of the right lateral geniculate body, where axons of a new neuron start as optic radiation, which ends in the right visual center of the right occipital lobe. The pupillary fibers leave the distal part of the optic tract to reach the Edinger–Westphal nucleus of the third cranial nerve after decussation. That explains why in a lesion of the visual pathway above the level of the lateral geniculate body pupils may be perfectly normal although the patient may have what is called cerebral blindness.

Thus together the optic nerve, optic chiasma, optic tract, lateral geniculate body, optic radiation, and occipital visual center are called the visual pathway. The defects in field of vision can be easily found out if nerve fibers in various parts of the visual pathway are known.

2.12 The Vitreous

A gelatinous body, forming two-thirds of the volume and weight of the eye, is found in the vitreous chamber, the space bounded by the lens, retina, and optic disc. The outer surface of vitreous, the hyaloid membrane, is normally in contact with the posterior lens capsule, the zonular fibers, the pars plana epithelium, the retina, and the optic nerve head. Its base has a firm attachment throughout life to the pars plana epithelium and the retina, just behind the ora serrata. Its attachment to the lens capsule and the optic nerve head is firm only in early life. It is made up of 99 % water and the remaining 1 % includes two components, collagen and hyaluronan, which confer a gel-like form and consistency because of their ability to bind large volumes of water. The optic disc normally has a central depression called the physiologic cup, which becomes very deep in glaucoma (Fig. 2.15).

2.13 Arrangement of Nerve Fibers in the Retina

The innermost layer of the retina is the nerve fiber layer. These nerve fibers are actually axons of the ganglion cells of the retina. The nerve fiber layer also has centrifugal fibers, fibers of Müller cells, neuroglial cells, and retinal vessels. The nerve fibers are arranged in bundles that run parallel to the surface of the retina. The bundles interweave with each other, forming a network in whose meshes are the processes of Müller and other glial cells. All the fibers converge toward the optic disc. Those from the nasal side reach it without interruption; those from the temporal side do not pass through the macula but they have to arch over it. The fibers above the horizontal meridian pass above the macula and those below pass under it. Toward the lateral side of the macula is a sort of raphae from which the nerve fibers arise in a pennate manner (Fig. 2.16). Those just to the lateral side of the macula encircle this structure closely, while the more lateral ones arch over it above and below in ever-increasing arcs.

The fibers from the macula itself pass straight in toward the temporal side of the disc, constituting the papillomacular bundle.

The nerve fiber layer is thickest around the margins of the disc and differs in different quadrants. Directly laterally it is thinnest in the region of the papillomacular bundle. The thickest parts are the upper and lower medial quadrants. This explains why swelling of the optic disc (called papilloedema) is first visible in the thicker upper and lower medial quadrants and last in the directly lateral part. From the disc, the nerve fiber layer becomes thinner when going toward the periphery and near the ora serrata, which is the thinnest layer. At the bottom of the fovea, it

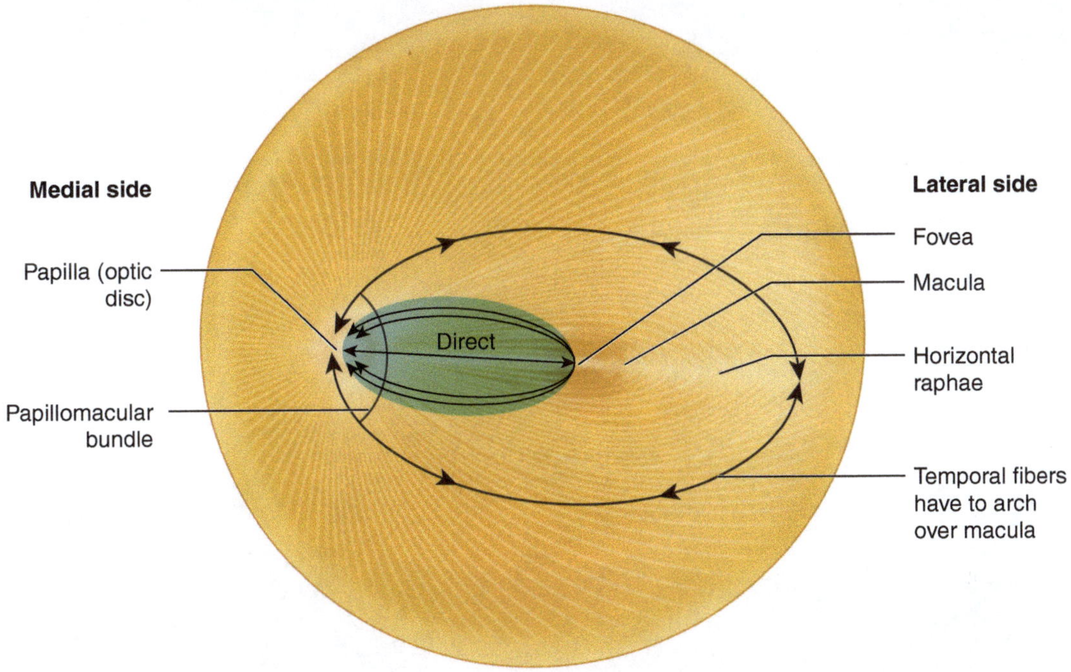

Fig. 2.16 Arrangement of retinal nerve fibers (*schematic*)

appears to disappear entirely. The nerve fibers are nonmodulated inside the eye. Individual fibers are separated by neuroglial processes.

The retinal blood vessels are found mainly in the nerve fiber layer, but they may also lie in the ganglion cell layer. They do not project on the inner surface of the retina. At the junction of the retina and vitreous is a membrane that forms both the inner limit of the retina and the outer boundary of the vitreous. There are no fibers of Müller at the optic disc, which has an important bearing on the swelling of the disc. There are no visual cells at the optic disc, which projects in the field of vision as a blind spot. It is important to remember that the retina is the only place in the human body where blood vessels of the arteriolar size can be directly seen by a direct ophthalmoscope, especially the changes in their walls. Since retinal circulation is an offshoot of cerebral circulation, the condition of the cerebral blood vessels can be guessed.

Suggested Reading

Parsons' Diseases of the Eye, edn 16. Revised by Miller SJH. London: Churchill Livingstone; 1979. pp. 1–14; 545–60

Riordon-Eva P, Cunningham E (2011) Vaughan and Asbury's general ophthalmology, edn 18. McGraw Hill, New York, pp 1–26

Trevor-Roper PD (1974) The eye and its disorders. Blackwell Scientific Publications, London, pp 26–44, 323–34

Wolff's E (1976) Anatomy of the eye and orbit, edn 7. Revised by Warwick R. London: HK Lewis; 1976; pp. 1–29; 30–180D

3 The Blood Supply to the Eyeball

Keywords
Proper refraction · Shape of eyeball · Circulation of the aqueous · Inflow · Outflow · Positive pressure · Microscopic outlet channels of aqueous veins · Collecting trunks · Dual blood supply in retina like brain

3.1 Shape of the Eyeball and Circulation of the Aqueous

For proper refraction to occur at the cornea, it is essential that the walls of the eyeball are kept tight by having a positive pressure inside the eyeball. This positive pressure is achieved by circulation of a liquid inside the eye called the aqueous humor. This liquid is chiefly secreted from blood in the posterior chamber from the fenestrated capillaries of the ciliary processes. The aqueous then travels to the anterior chamber of the eye via the pupil. Then it reaches the periphery of the anterior chamber (called the angle of the anterior chamber), which has microscopic outlet channels for the aqueous, namely, the trabecular meshwork, the canal of Schlemm, 12 aqueous veins, and 30 collector channels. The aqueous is drained through these microscopic outlet channels to ultimately reach the episcleral venous plexus. This inflow and outflow of the aqueous from and to the blood creates a positive pressure inside the eyeball, which maintains its shape so that refraction at the cornea occurs normally, and the eyeball is not indented by atmospheric pressure.

3.2 The Blood Supply to the Eye and Related Structures

The principal arterial supply of the orbit and the eyeball derives from the ophthalmic artery, the first major branch of the intracranial portion of the internal carotid artery. This branch passes beneath the optic nerve and accompanies it through the optic canal into the orbit. The first intraorbital branch of the ophthalmic artery is the central retinal artery (CRA), which enters the optic nerve about 15 mm behind the globe. Other branches of the ophthalmic artery include the lacrimal, which supplies the lacrimal gland and the upper lid, muscular branches to various muscles of the orbit, long and short posterior ciliary arteries, medial palpebral arteries to both eyelids, and supraorbital and supratrochlear arteries. The short ciliary arteries supply the choroid and the optic disc. The two long posterior ciliary arteries supply the

M.W. Ansari and A. Nadeem, *Atlas of Ocular Anatomy*, DOI 10.1007/978-3-319-42781-2_3

ciliary body and anastomose with each other and with the anterior ciliary arteries to form the major arterial circle of the iris. They supply the anterior sclera, episclera, limbus, and conjunctiva and contribute to the formation of the major arterial circle of the iris. The central retinal artery, the first intraorbital branch of the ophthalmic artery is an end artery like the arteries of the brain because from an embryologic point of view the retina is a part of the brain. It has a dual supply like the brain, the outer one-third by choriocapillaris and the inner two-thirds by branches of the CRA. Since it is an end artery, its occlusion will result in a total loss of vision.

The blood supply to the optic nerve and the visual pathways is through the branches of the ophthalmic and internal carotid arteries via their pial plexuses. The axial part of the intracranial optic nerve has fewer vessels supplying more nerve fibers because septal meshes are wider there. Any disturbance of the blood supply to that part can affect the papillomacular fibers.

3.3 Circulation of the Aqueous and the Shape of the Eyeball

We have already mentioned in Chap. 2 that the eyeball must maintain its shape so that refraction can occur properly at the corneal surface, that is, the walls of the eyeball have to be tight so that atmospheric pressure does not indent it. That purpose is achieved by circulation of a fluid inside the eye called the aqueous humor. This fluid is chiefly secreted in the posterior chamber from the fenestrated capillaries of the ciliary processes (Fig. 3.1). The aqueous travels from the posterior to the anterior chamber through the pupil. It then reaches the periphery of the anterior chamber, called the angle, which has microscopic outlet channels of the aqueous (the trabecular meshwork, the canal of Schlemm, and the aqueous veins and collecting trunks that drain the aqueous finally into the episcleral venous plexus). Thus inflow from blood occurs in the posterior chamber, and outflow is achieved

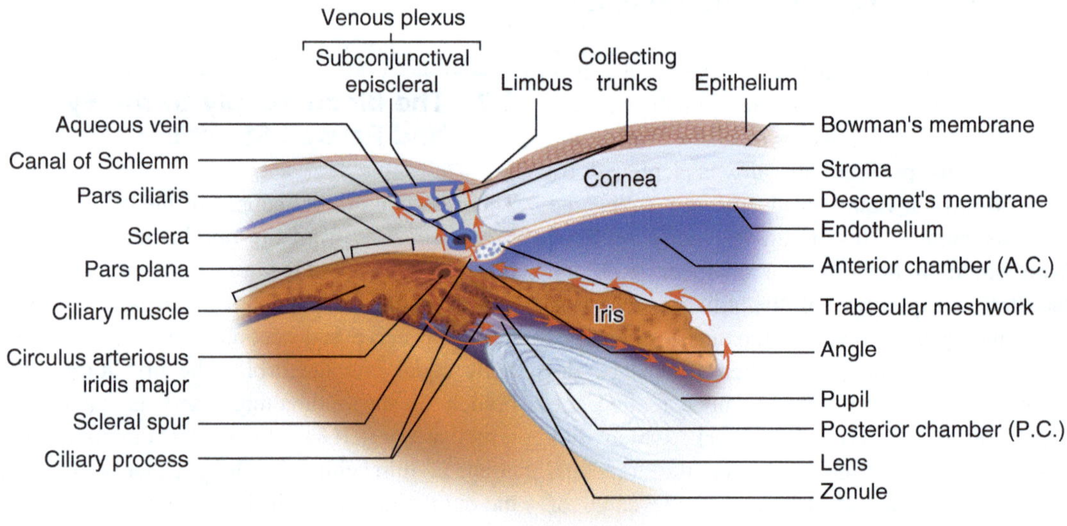

Fig. 3.1 Circulation of the aqueous (*arrows*) produced chiefly in the posterior chamber: pupil → anterior chamber → angle of the anterior chamber → trabecular meshwork → canal of Schlemm → aqueous veins (12), 30 collecting trunks → episcleral venous plexus. Note the microscopic outlet channels of the aqueous

through the microscopic outlet channels. A balance between inflow and outflow maintains a positive pressure inside the eye called the intraocular pressure, which prevents indentation of the eyeball by atmospheric pressure. The small arrows in Fig. 3.1 illustrate this circulation. A fraction (20 %) of the aqueous is drained to the suprachoroidal space through the ciliary body. it is called the uveoscleral outflow.

3.4 The Blood Supply to the Eyeball: The Ophthalmic Artery

Embryologically the retina is a part of the brain, and therefore they have many common denominators such as a dual blood supply. The retina has one-third peripheral blood supply from the choriocapillaris and two-thirds inner blood supply from the branches of the central retinal artery, which is an end artery. The blood supply to the eye and its adnexa is from the ophthalmic artery, which is the first intracranial branch of the internal carotid artery that supplies the brain. That explains why it is more prone to embolism.

3.5 The Ophthalmic Artery

The ophthalmic artery arises inside the skull as the first intracranial branch of the internal carotid artery from its fifth bend, as soon as it emerges from the roof of the cavernous sinus. At its origin, the ophthalmic artery is medial to the anterior clinoid process and inferior to the optic nerve (Figs. 3.2, 3.3 and 3.4). It passes through the optic canal within the sheath of the optic nerve,

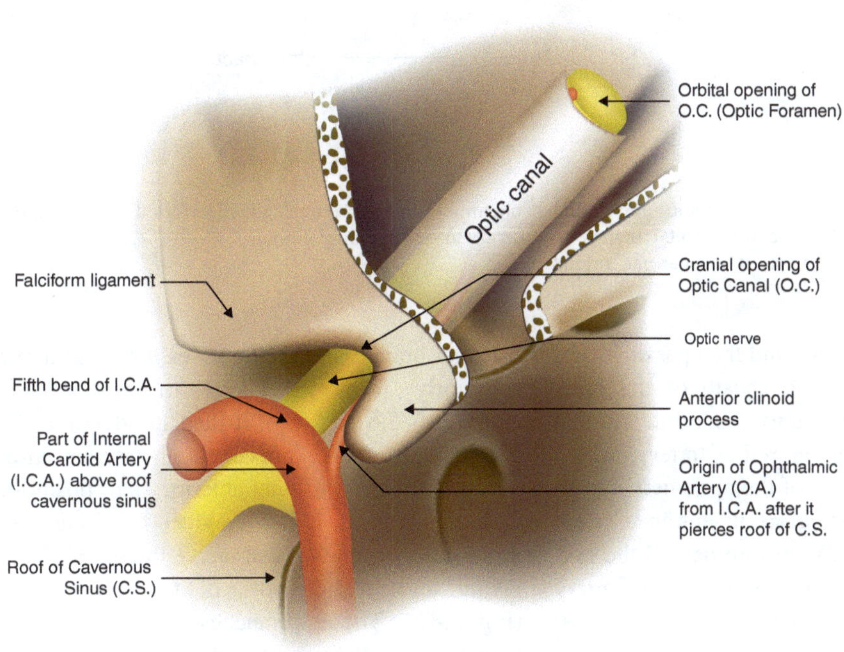

Fig. 3.2 Origin of the ophthalmic artery from the internal carotid artery as soon as it emerges from the roof of the cavernous sinus from its fifth bend

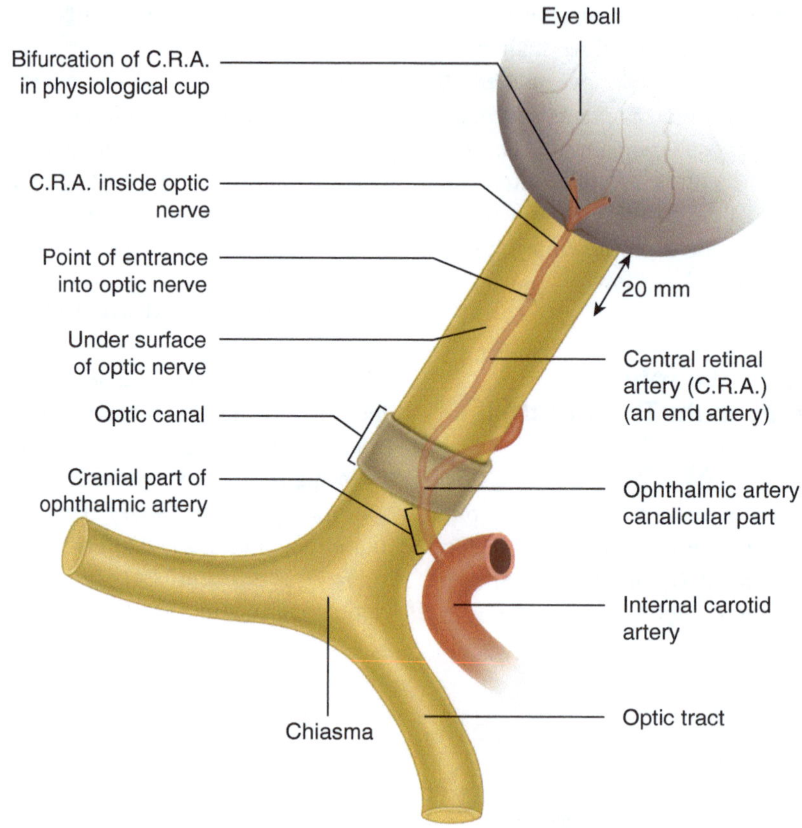

Fig. 3.3 Highly diagrammatic ophthalmic artery and CRA. Note that the CRA is the first orbital branch of the ophthalmic artery as soon as it reaches the orbital opening of the optic canal, and therefore it becomes more prone to occlusion by emboli

at first inferior to the nerve and then passing to its lateral side. It pierces the sheath of the optic nerve near its entrance into the orbit. At that point, it gives the central retinal artery, which travels outside the sheath of the optic nerve on its undersurface (Fig. 3.5) before its entrance inside the optic nerve about 15–20 mm behind the eye. Because from here onward the ophthalmic artery has to supply structures inside the cone of muscles, it enters the cone on the lateral side of the optic nerve and medial to the ciliary ganglion. Once inside the cone of muscles, it gives its lacrimal branch and then crosses the optic nerve superiorly from its lateral to medial side (Figs. 3.4, 3.6 and 3.7). It then divides into the supratrochlear and dorsal nasal arteries after giving branches to the anterior and posterior ethmoidal canals. It also provides two long posterior ciliary branches that pierce the sclera to the lateral and medial sides of the optic nerve (Fig. 3.8). These branches move forward between the sclera and the choroid to supply the ciliary body and then anastomose with the anterior ciliary arteries to form the circulus arteriosus

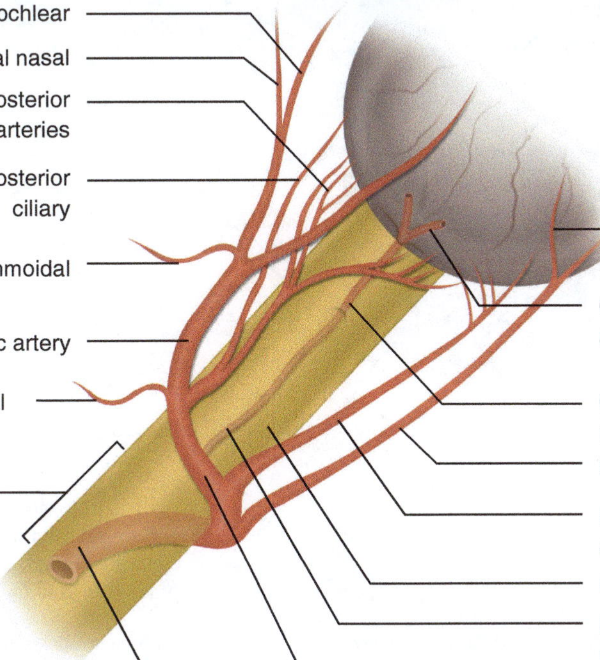

Fig. 3.4 The orbital canalicular part of the ophthalmic artery, which is the first intracranial branch of the internal carotid artery. Note its three parts: cranial, inside optic canal, and orbital. The orbital part crosses the optic nerve superiorly. Note that the CRA is an end artery like the brain arteries because the retina is a part of the brain and has a dual blood supply

iridis major (Fig. 3.9), which supplies the iris and the ciliary processes. Ten to twenty short posterior ciliary branches run forward and surround the optic nerve near the lamina cribrosa and pierce the eyeball to supply the optic disc, forming the circle of Zinn.

3.6 The Central Retinal Artery

Figures 3.4 and 3.6 confirms the origin of the central retinal artery (CRA) from the ophthalmic artery, close to the optic foramen. It runs a wavy

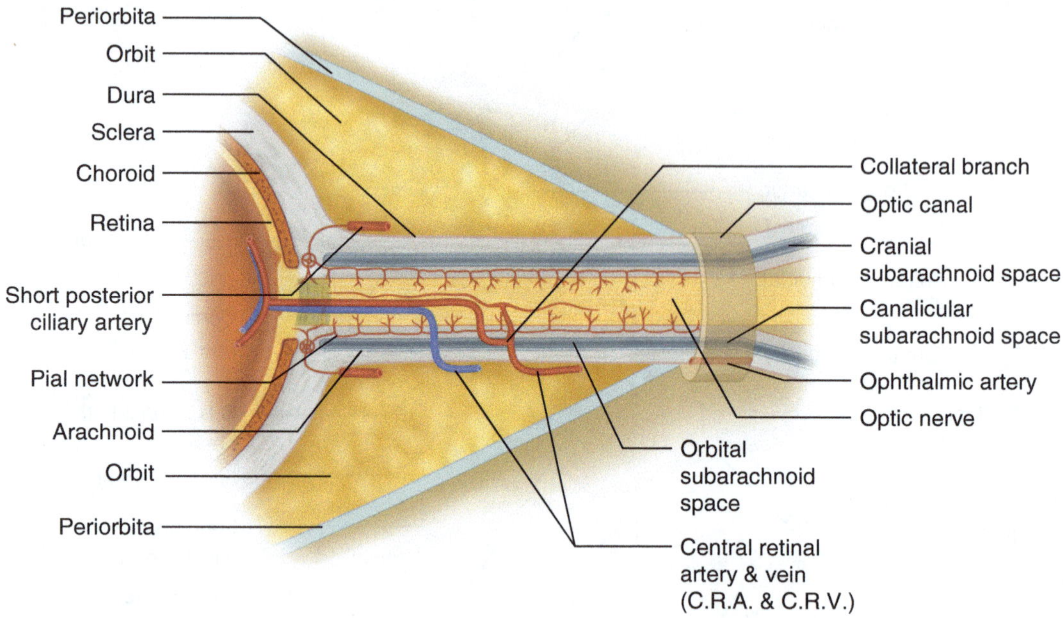

Fig. 3.5 The blood supply to the optic nerve. Note the three sheaths of optic nerve: the dura, the arachnoid, and the pia. Note also that the subarachnoid spaces of the cranial, canalicular, and orbital parts of the optic nerve are continuous with each other. The CRA takes (1–5) bends before reaching the optic disc. The sclera is continuous with the dura of the optic nerve on the back of the eye

course forward on the undersurface of the optic nerve, outside but adherent to the dural sheath to about 15–20 mm behind the eye, where it enters the optic nerve, piercing the dura and the arachnoid (see Fig. 3.5). Upon reaching the subarachnoid space it bends forward and after a very short course it again bends upward at a right angle, invaginating the pia. At the center of the optic nerve, it bends forward in company with its vein. It passes anteriorly to the lamina cribrosa, piercing it and then, climbing the nasal side of the physiologic cup, it enters the retina. Thus it makes five bends (including its bifurcation) from the entry into the optic nerve to the physiologic cup.

3.7 The Blood Supply of the Visual Pathway

The network is supplied by arteries that probably anastomose slightly in the network but not before they reach it. The vessels passing into the nerve take with them a coat of pia and a covering of glia that constitute the septa. The vessels are always separated from the nerve tissue by perivascular glia. As the vessels pass into the nerve in the septa, they divide dichotomously and send branches anteriorly and posteriorly.

(A) Intracranial part: pial network supplied by branches from the ophthalmic superior

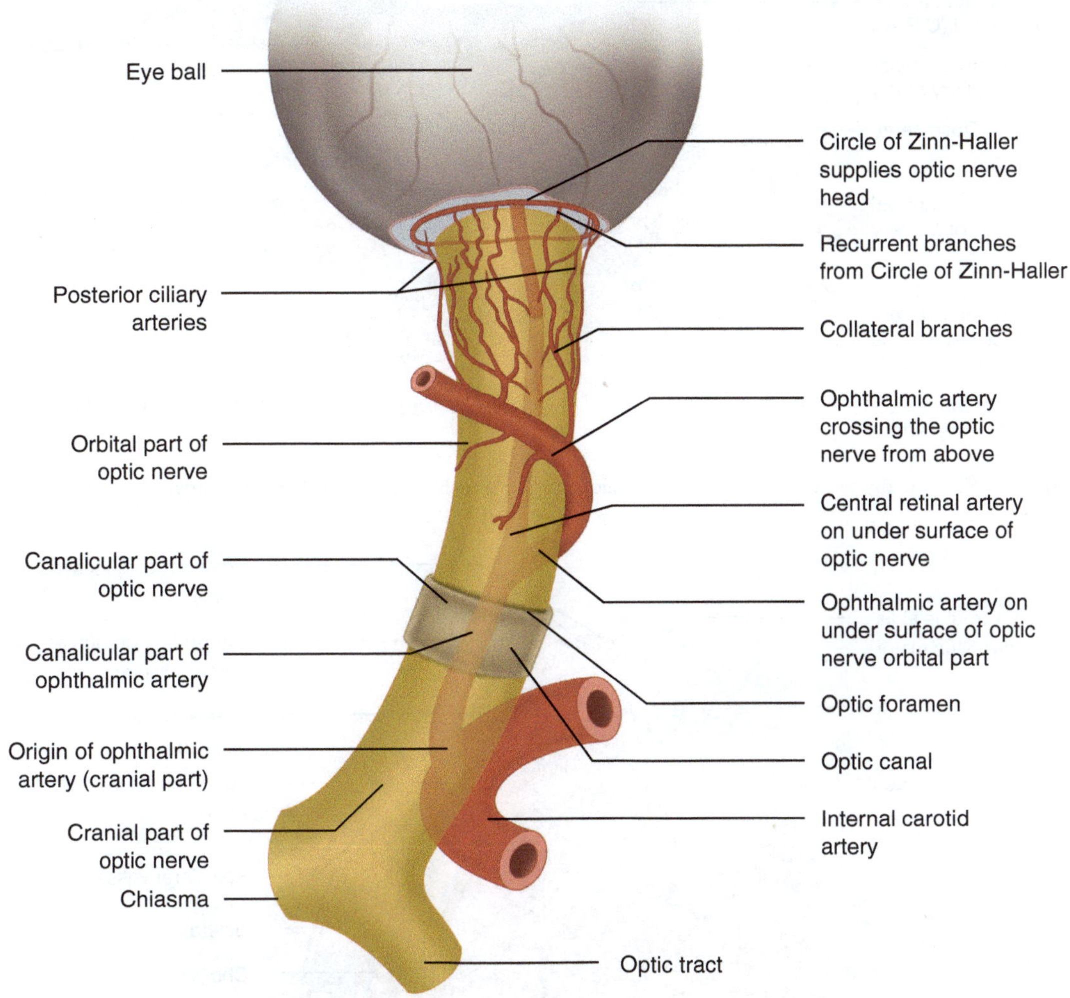

Fig. 3.6 The blood supply to various parts of the optic nerve: cranial from the internal carotid artery, canalicular from the ophthalmic artery, orbital from the collateral branches of the ophthalmic artery and the CRA, short and long ciliary arteries and the CRA, the optic disc from the circle of Zinn

hypophyseal, anterior cerebral, and internal carotid arteries. In the axial area, the septa are so thin that smaller vessels supply a large number of nerve fibers. This explains why any disturbance of the blood supply in that area will affect the papillomacular bundle.

(B) Parts other than the intracranial part such as the intracanicular part are supplied by branches of the ophthalmic artery. The orbital part of the optic nerve is supplied by two groups of vessels: (1) those that pierce the dura behind the entrance of the CRA, six to a dozen branches from the ophthalmic artery and its branches. (2) Where the CRA enters the nerve it supplies a collateral branch that passes backward toward the optic foramen (see Fig. 3.5).

3.8 Venous Drainage of the Orbit

Figure 3.10 shows superior and inferior ophthalmic veins (SOVIOVs). Their tributaries have no valves and are markedly tortuous. They communicate with the veins of the face, the pterygoid

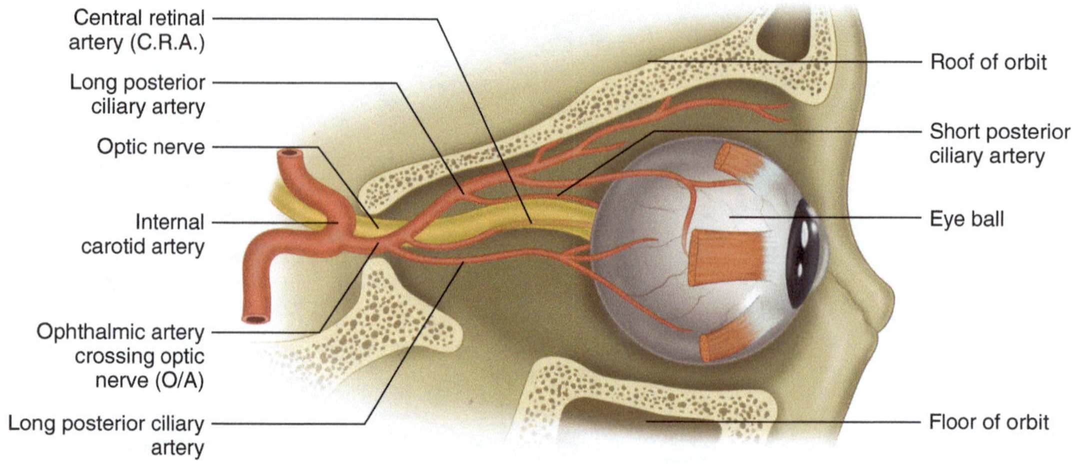

Fig. 3.7 Blood supply to the eye by the ophthalmic artery, a branch of the internal carotid artery

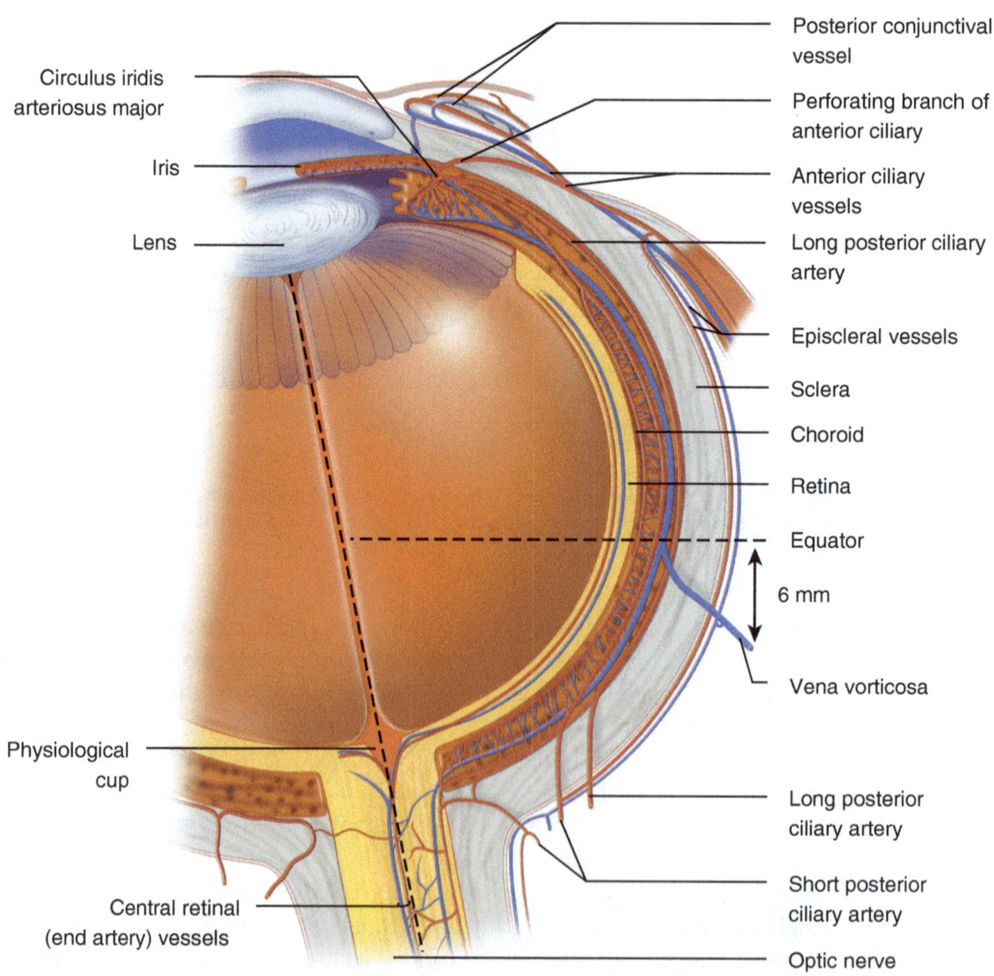

Fig. 3.8 The blood supply to the eye: branches of the ophthalmic artery

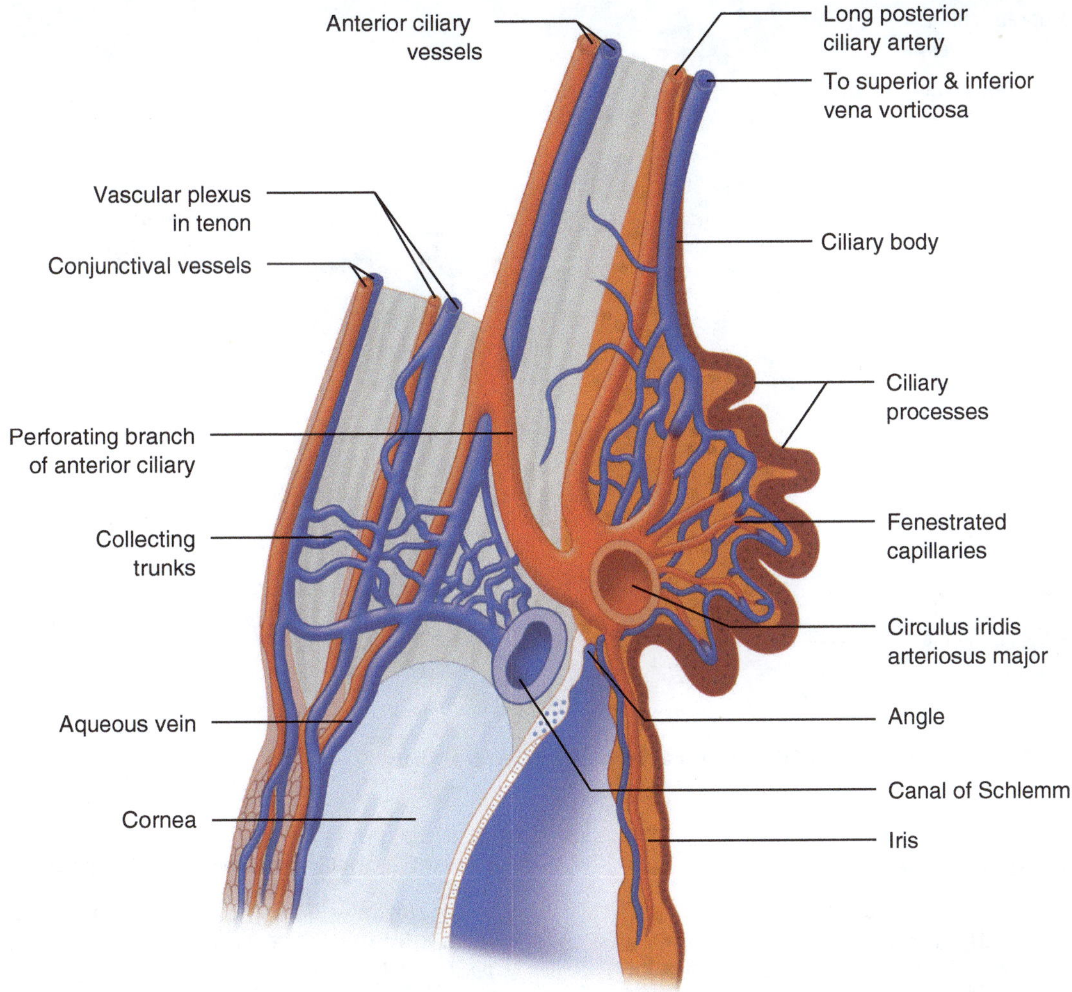

Fig. 3.9 Blood vessels to the anterior segment of the eye

plexus, and the nose. They drain directly into the cavernous sinus.

Tributaries include the inferior ophthalmic, the anterior and posterior ethmoidal, the lacrimal, the central vein of the retina, the anterior and posterior ciliary, and four vertex veins.

Any neglected periorbital septic spot can give rise to thrombosis of the cavernous sinus.

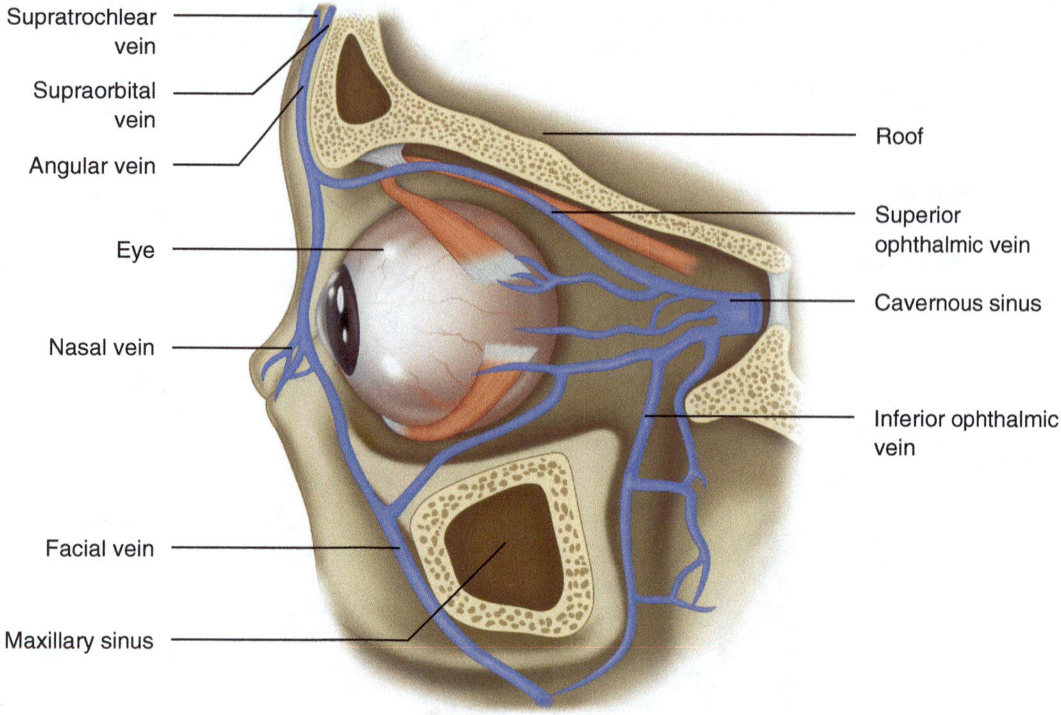

Fig. 3.10 Venous drainage of the ophthalmic orbit to the cavernous sinus

Suggested Reading

Parsons' Diseases of the Eye, edn 16. Revised by Miller SJH. London: Churchill Livingstone; 1979. pp. 1–14; 545–60

Riordon-Eva P, Cunningham E (2011) Vaughan and Asbury's general ophthalmology, edn 18. McGraw Hill, New York, pp 1–26

Trevor-Roper PD (1974) The eye and its disorders. Blackwell Scientific Publications, London, pp 26–44, 323–34

Wolff's E (1976) Anatomy of the Eye and Orbit, edn 7. Revised by Warwick R. London: HK Lewis; 1976; pp. 1–29; 30–180D

Extraocular and Intraocular Muscles 4

Keywords
Extraocular and intraocular muscles · Four recti · Two obliques · Levator palpebrae superioris · Orbicularis oculi · Blinking reflex · Unstriated involuntary muscles · Cervical sympathetic · Primary action · Subsidiary action · Neuromuscular control

4.1 Extraocular Muscles

The extraocular muscles include the orbicularis oculi, levator palpebrae superioris, four rectus and two oblique muscles, totaling eight. The four rectus muscles have a common origin from an oval ring at the apex of the orbit called the annulus of Zinn, which encompasses the optic foramen at the back and the wider medial part of the superior orbital fissure. The ring has an upper and a lower tendon. The upper part supplies origin to part of the medial and lateral recti and the whole of the superior rectus. The lower part provides origin to part of the lateral and medial recti and the whole of the inferior rectus. The origins of the superior and medial recti are more closely related to the neighboring sheaths of the optic nerve. This explains the occurrence of pain during extreme movements of the globe in retrobulbar optic neuritis. Their length is about 40 mm, the superior rectus being the longest and the inferior rectus being the shortest. They extend anteriorly, close to the walls of the orbit. Their edges are interconnected by an intermuscular septum, forming a cone of muscles that divides the orbital cavity into a peripheral and a central space. This cone encloses the optic nerve and vessels and the nerves supplying them. Naturally, the nerves and vessels that supply structures outside the cone pass outside the annulus, such as the lacrimal, frontal, and trochlear nerves. The nerves that supply structures inside the cone of muscles pass through this annulus, such as the two divisions of the third cranial nerve, the nasociliary nerve, and the sixth cranial nerve. The four recti proceed anteriorly toward the limbus to be inserted at variable distances from the limbus in front of the equator of the eye. Each muscle has one main action and subsidiary actions. The superior rectus elevates the eye to its maximum degree when an eye is abducted 25°. Its subsidiary actions are adduction and intorsion. The inferior rectus is a depressor in the abducted position, its subsidiary actions being adduction and intorsion. The lateral rectus is an

M.W. Ansari and A. Nadeem, *Atlas of Ocular Anatomy*, DOI 10.1007/978-3-319-42781-2_4

abductor, and the medial rectus is an adductor. The superior oblique muscle, the longest and thinnest, arises from above and medial to the optic foramen and passes forward between the roof and medial wall to the trochlear pulley. After passing through the trochlea, the tendon takes a bend backward, downward, and laterally by an angle of 55° to the visual axis and gets inserted obliquely in the posterosuperior quadrant by a tendon that is convex backward and laterally. The main action of the superior oblique is depression, which is at its maximum when the eye is adducted. The subsidiary actions, abduction and intorsion, increase as the eye is abducted. The inferior oblique muscle is the only one arising from the floor of the orbit just lateral to the opening of the nasolacrimal duct. It passes laterally and backward at an angle of 55° to the visual axis, roughly parallel with the tendon of the superior oblique muscle and is inserted at the back and lateral portions of the globe below the horizontal meridian. The line of insertion is oblique and convex upward. The main action of the inferior oblique muscle is elevation, which is at its maximum in the adducted position. The subsidiary actions are abduction and extorsion, which are greater in the abducted position.

The levator palpebrae superioris (LPS) arises from the undersurface of the lesser wing of the sphenoid, blends with the origin of the superior rectus, and passes forward on the superior rectus, reaching a few millimeters in front of the equator. It divides into a deep and a superficial part. The deep part goes to the upper border of the upper tarsus, and the superficial part becomes a thin triangular structure called its aponeurosis. The base of the triangle is inserted into the undersurface of the skin of the upper lid after passing through the subcutaneous orbicularis oculi muscle there. The lateral end of the aponeurosis divides the main lacrimal gland into a palpebral and an orbital part, to be inserted in the Whitnal tubercle, and its medial end gets attached to medial palpebral ligament. The levator raises the upper eyelid, thus uncovering the cornea and a portion of the sclera and deepens the superior palpebral fold. Its antagonist is the palpebral portion of the orbicularis.

The nerve supply of the superior rectus (SR), the inferior rectus (IR), medial rectus (MR), inferior oblique (IO), and levator are supplied by the third cranial nerve and the lateral rectus by the sixth cranial nerve. The orbicularis oculi is supplied by the seventh cranial nerve. Its palpebral part arises from the medial palpebral ligament and is inserted into the lateral palpebral raphae. The orbital part arises from the medial side of the orbit and the medial palpebral ligament. The peripheral fibers sweep across the orbital margin in a series of concentric loops.

The actions of the orbicularis oculi are as follows. The palpebral part is used in the involuntary act of blinking, and the orbital part is used to close the lids forcefully. There are three intraocular muscles. Two are in the iris—the sphincter pupillae and the dilator pupillae. The third is in the ciliary body and is called the ciliary muscle. Its longitudinal fibers increase the outflow of the aqueous, and the circular fibers help in producing accommodation.

4.2 Eight Extraocular Muscles

There are eight extraocular muscles in each eye, including the levator and orbicularis oculi as well as four recti and two obliques. Since they are derived from branchial arches, they are more highly differentiated than any other muscles in the body. Instead of being grouped together in bundles and separated by dense connective tissue, their fibers are loosely united. Each eye muscle receives a nerve supplied according to its size. These nerves contain more nerve fibers than other muscular nerves. Each motor neuron supplies relatively few muscle fibers. The muscle fibers are of two types: (1) thin fibers, also called slow fibers, are distributed peripherally and have grapelike motor terminals; (2) twitch fibers are wider than slow fibers and have plaque-like motor terminals. Four of the rectus muscles (superior, inferior, medial, and lateral) have their origin from an oval tendinous ring at the apex of the orbit called the annulus of Zinn, which surrounds the back of the optic foramen and the wider medial part of the superior

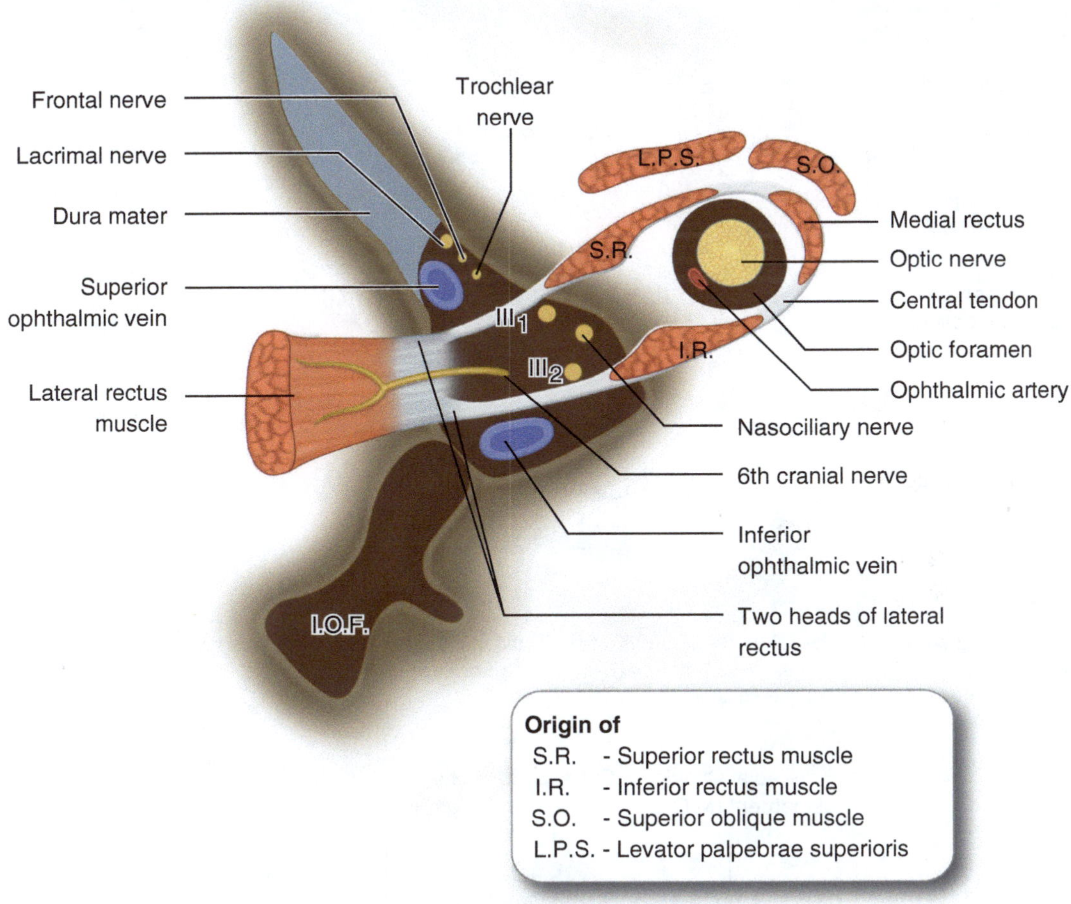

Fig. 4.1 Superior orbital fissure (SOF) and central tendon of Zinn and structures that are passing through

orbital fissure in front (Figs. 4.1 and 4.2). It is thickened on its inner side, and its upper and lower parts give origin to part of the lateral and medial recti and all of the inferior rectus muscles. The origins of the superior and medial recti are more closely attached to the sheath of the optic nerve. This explains why extreme movements of globe are painful in retrobulbar optic neuritis.

The four recti are about 40 mm long. The superior rectus is the longest, and the inferior rectus is the shortest. They extend anteriorly close to the walls of the orbit. Their edges are connected to each other by the intermuscular septum, which forms a cone of muscles dividing the orbital cavity into a peripheral and a central space (Fig. 4.3). They are inserted into the sclera, well anterior to the equator of eye by tendons of different widths and at different distances from the limbus (Fig. 4.4). The sclera is very thin at the insertions of the rectus muscles. This explains why traumatic rupture of the globe is common in this area.

Figure 4.1 shows that the superior rectus muscle arises from the upper part of the annulus of Zinn above and lateral to the optic foramen and also from the neighboring sheath of the optic nerve. On the one hand, this origin lies in the angle formed by the splitting of the dura that lines the optic canal to form the orbital periosteum (periorbita), and on the other hand it lies in the dural covering of the nerve. It is below the levator muscle and is continuous on the medial side with the medial rectus and on the lateral with

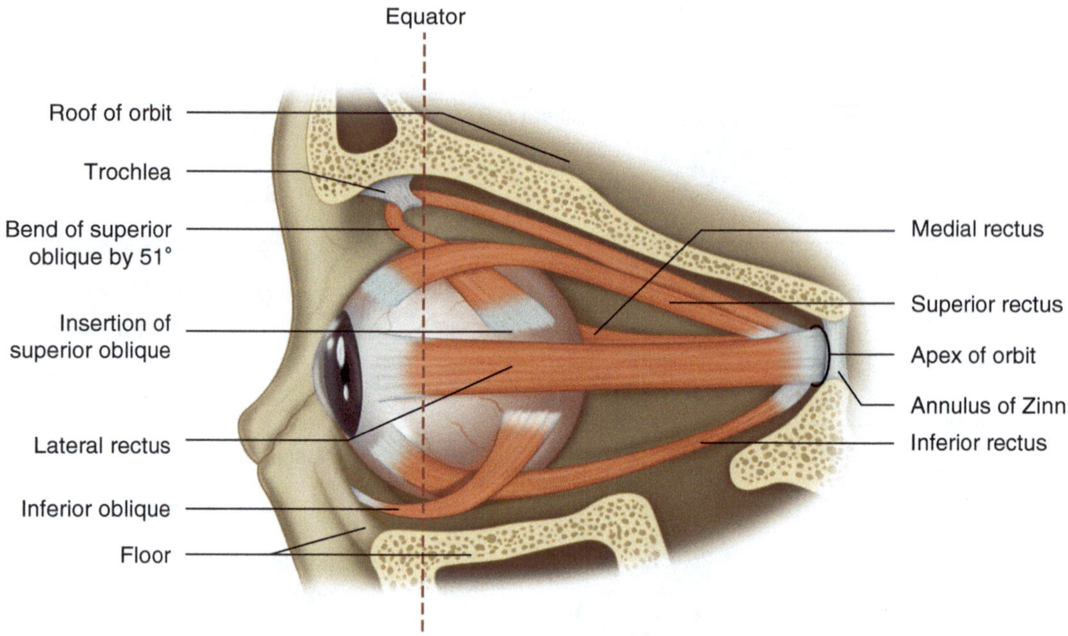

Fig. 4.2 Extraocular muscles. Note the insertion of the oblique muscles behind the equator and the insertion of the rectus muscles in front of the equator. The origin of the inferior oblique muscle from the floor of the orbit can also be seen

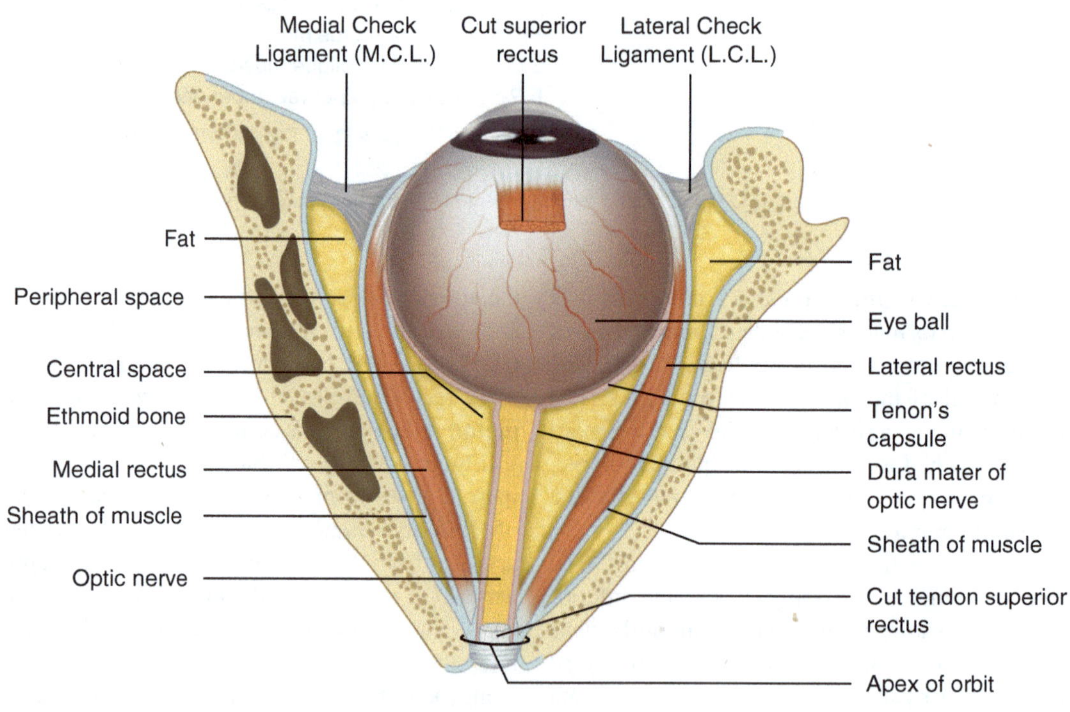

Fig. 4.3 **a** Tenon capsule (from limbus to optic nerve). **b** Cone of muscles

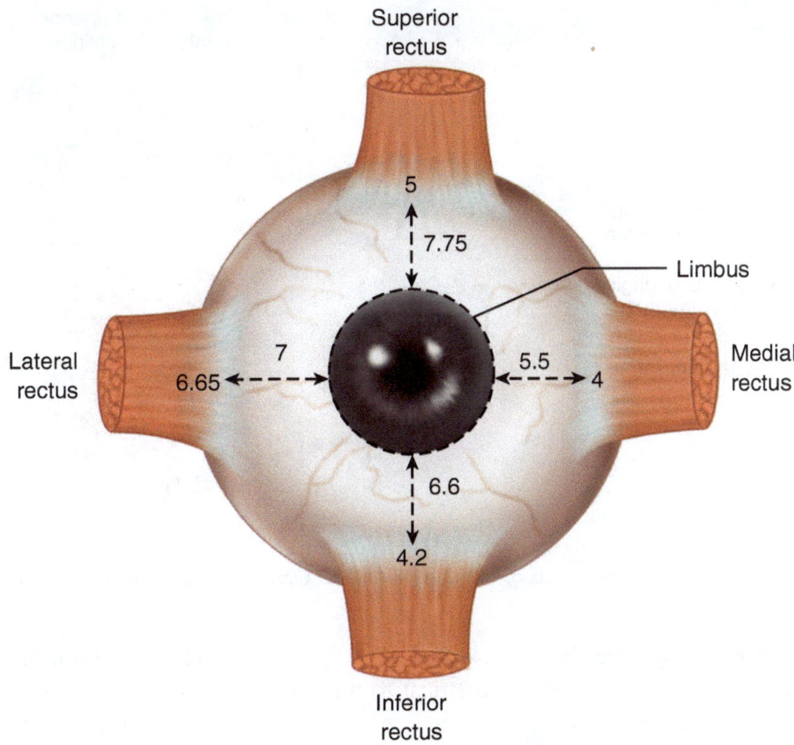

Fig. 4.4 Lines of insertion, rectus muscles

the lateral rectus. The muscle passes forward and laterally beneath the levator muscle at an angle of 23–25° to the visual line, pierces the Tenon capsule, and is inserted into the sclera 7.75 mm from the cornea (see Fig. 4.4).

The nerve to the superior rectus is the superior division of the third cranial nerve, which enters the muscle at the junction of its middle and posterior thirds. The blood supply is from the lateral muscular branch of the ophthalmic artery.

The primary action of the superior rectus muscle is elevation, which is at its maximum when the eye is abducted 23° in the line of its insertion. Subsidiary actions are adduction and intorsion, which increase as the eye is adducted or rotated medially.

The inferior rectus muscle arises wholly from the lower tendon of the annulus of Zinn (see Fig. 4.1) below the optic foramen, passes forward and laterally along the floor of the orbit at an angle of 23–25° with the visual line, and is inserted into the sclera 6.5 mm from the limbus (see Fig. 4.4). It sends a fascial expansion of its sheath to the lower border of the lower tarsus. The inferior oblique muscle crosses below the inferior rectus, where the sheaths of two muscles are united. At about the junction of its posterior and middle third on its ocular surface, the inferior oblique is supplied by the inferior division of the oculomotor nerve. Its blood supply is from the medial muscular branch of the ophthalmic artery.

The primary action of the inferior rectus muscle is depression, which is at its maximum when the eyeball is abducted 23–25°; subsidiary actions are adduction and extorsion, which increase as the eye is turned in.

The medial rectus has its origin in the medial side and below the optic foramen from both parts of the common tendon (upper and lower) and also from the sheath of the optic nerve. That explains the occurrence of pain in its extreme movement in retrobulbar optic neuritis. It passes forward along the medial wall of the orbit and is inserted into the sclera 5.5 mm from the limbus

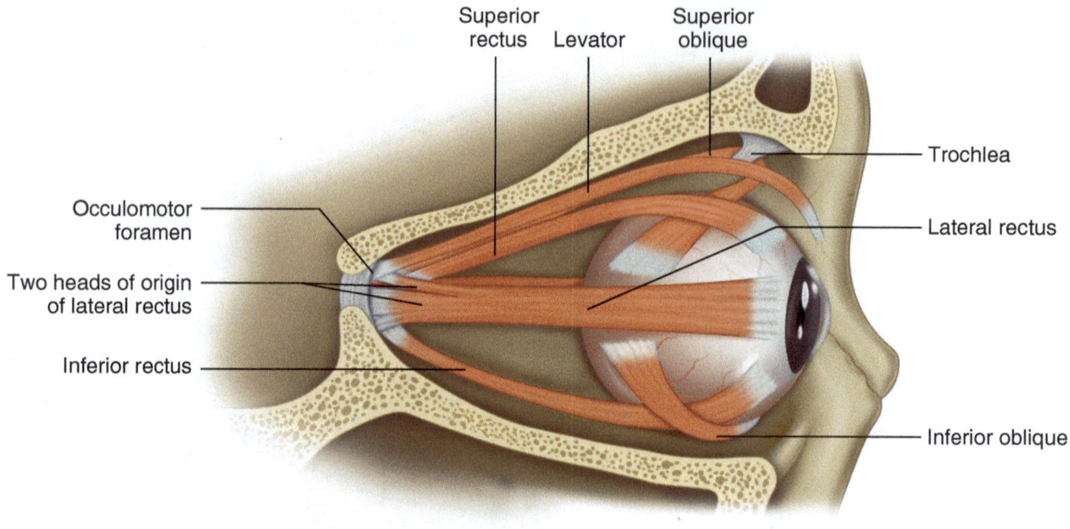

Fig. 4.5 Note the two heads of origin of the lateral rectus muscle forming the oculomotor foramen

(see Fig. 4.4). The line of insertion is 10.3 mm long and is straight and symmetrical to the horizontal meridian.

Its primary action is pure adduction. Both muscles act together in convergence when the eyeball is in primary position.

The lateral rectus muscle arises from both the upper and lower parts of the common tendon, from those portions that bridge the superior orbital fissure (Fig. 4.5). This origin is continuous and is strengthened by its attachment to the spina recti lateralis on the greater wing of the sphenoid. The origin thus looks like a letter U placed so that the opening looks toward the optic foramen; the limbs of the U are referred to as the upper and lower heads of the muscle. A small interval is between the origin of the lateral rectus and that portion of the lesser wing that separates the optic nerve from the medial portion of the superior orbital fissure. The structures that pass through this interval are described as passing between the two heads of the lateral rectus within the cone of muscles or the annulus of Zinn; they supply structures within the muscle cone. They are from above downward the upper division of the third cranial nerve, the nasociliary nerve, and the lower division of the third cranial nerve. The sixth cranial nerve is at a level between the two divisions. Outside the annulus and above the upper head of the lateral rectus are the lacrimal, frontal, and trochlear nerves. They pass outside the annulus because they will supply structures outside the cone of muscles. The inferior ophthalmic vein may pass below and outside the annulus. The thin superolateral part of the superior orbital fissure is covered by dense fibrous tissue, and nothing passes through it. The nerve supply of the lateral rectus is from the sixth cranial nerve which enters the muscle at its middle on its medial surface.

4.3 The Fascial Hammock

The eyeball is held in position in the front part of the orbit by lateral and medial check ligaments on its sides and inferiorly by the ligament of Lockwood. The lateral and medial check ligaments actually are fascial extensions of insertions of lateral and medial recti on the sclera, extending to the lateral and medial walls of the orbit. Inferiorly, the eyeball is supported by a regional

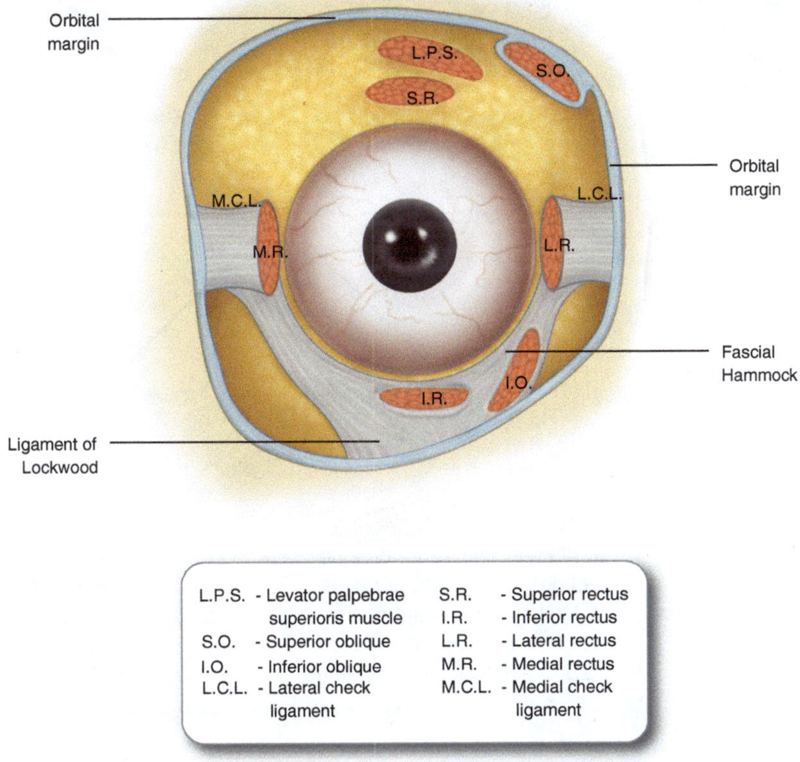

Fig. 4.6 Fascial hammock on which the eyeball rests in the orbit formed by the medial check ligament (MCL) and the lateral check ligament (LCL) on the sides and inferiorly by the ligament of Lockwood (L/L)

thickening of the orbital fascia called the ligament of Lockwood. Actually, the two check ligaments join the ligament of Lockwood to form a fascial hammock (Fig. 4.6) on which the eyeball rests, enveloped by the Tenon capsule from limbus to optic nerve. The inner surface of the Tenon capsule is very smooth and acts like a capsule in which the eyeball rotates in various directions along its center of rotation, without being displaced in space. Thus whenever we talk of ocular motility, we mean rotation of the eye around the center of rotation along three axes and elevation and depression along a horizontal axis passing through center of rotation. The adduction and abduction occur along a vertical axis. The intorsion, the movement of 12 o'clock of the cornea inward, and the extorsion, the turning out of 12 o'clock of cornea, both occur along an anteroposterior axis, again passing through the center of rotation of the eye, which almost corresponds to the center of the eyeball.

As elsewhere, the extraocular muscles do not work in isolation but in groups of two or more synergistically. Thus the visual axis can only be elevated vertically by the synergistic action of the superior rectus and the inferior oblique muscles, and depression must involve contraction of both the inferior rectus and the superior oblique muscles. To summarize, all 12 muscles are exquisitely controlled in a total group in all

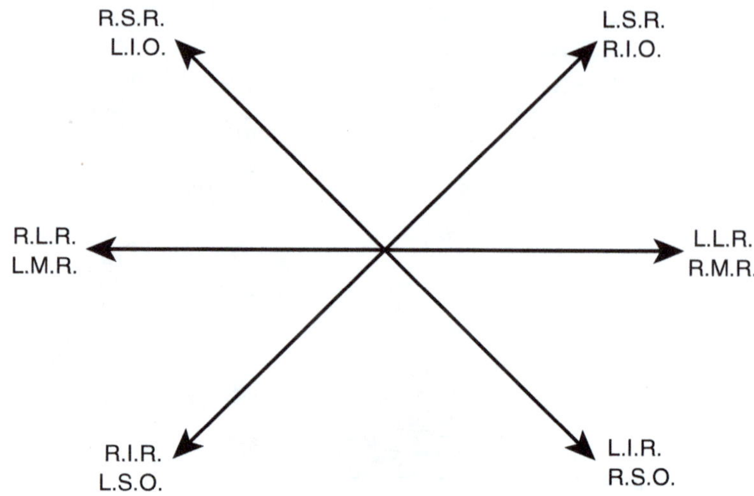

R.S.R. – Right superior rectus	L.S.R. – Left superior rectus
L.I.O. – Left inferior oblique	R.I.O. – Right inferior oblique
R.L.R. – Right lateral rectus	L.L.R. – Left lateral rectus
L.M.R. – Left medial rectus	R.M.R. – Right medial rectus
R.I.R. – Right inferior rectus	L.I.R. – Left inferior rectus
L.S.O. – Left superior oblique	R.S.O. – Right superior oblique

Fig. 4.7 The field of action of the ocular muscles in binocular movements. *LIO* Left Inferior oblique; *LIR* left inferior rectus; *LLR* left lateral rectus; *LMR* left medial rectus; *LSO* left superior oblique; *LSR* left superior rectus; *RIO* right inferior oblique; *RIR* right inferior rectus; *RLR* right lateral rectus; *RMR* right medial rectus; *RSO* right superior oblique; *RSR* right superior rectus

ocular movements, with the constant feedback of vision to correct any departures leading toward diplopia. Thus the medial, superior, and inferior recti act as an adductor group, their abductor opponents being the lateral rectus and both oblique muscles. An example of synergists follows. On looking up and to the right, the right superior rectus and left inferior oblique muscles are primarily involved. The corresponding groups are shown in Fig. 4.7.

4.4 The Superior Oblique Muscle

Figures 4.1 and 4.2 show that the superior oblique muscle, the longest and thinnest muscle, arises above and medial to the optic foramen by a narrow tendon that partially overlaps the origin of the levator muscle. The fusiform muscle belly, more rounded than that of other extrinsic muscles, passes forward between the roof and the medial wall of the orbit to the trochlear pulley. The pulley is a U-shaped fibrocartilage closed above by fibrous tissue; it is attached to the spina trochlea, which is on the lower aspect of the frontal bone a few millimeters behind the orbital margin. Through the pulley the tendon is enclosed in a synovial sheath, beyond which a strong fibrous sheath accompanies the tendon to the eyeball. The muscle, which is about 1 cm behind the trochlea, gives place to a rounded tendon that passes through the pulley and then bends downward, backward, and laterally at an angle of 55°; it then pierces the Tenon capsule, passes under the superior rectus spreading out in a fan-shaped manner, and is inserted obliquely in the posterosuperior quadrant almost entirely lateral to the mid-vertical plane. The line of insertion is about 10.7 mm long and is convex backward and laterally (Fig. 4.8).

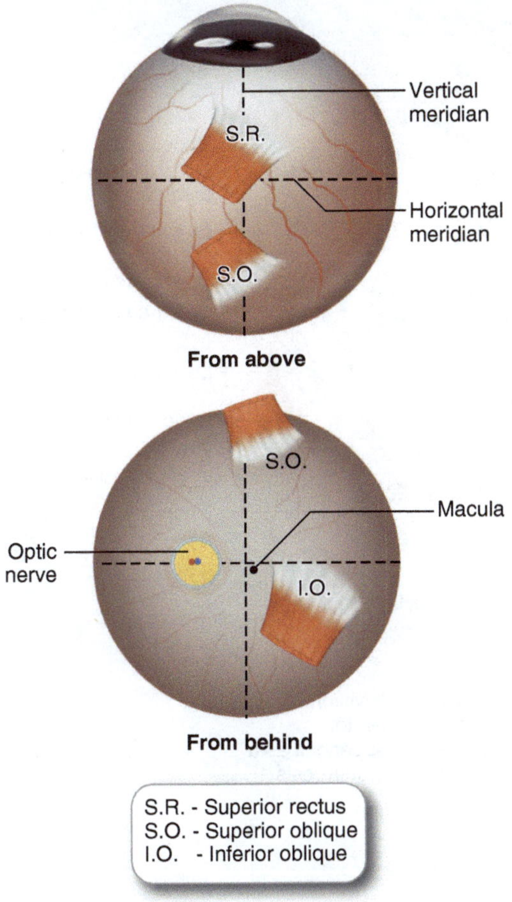

Fig. 4.8 Insertion lines of superior rectus (SR), superior oblique (SO), and inferior oblique (IO) muscles

4.4.1 Primary Action of the Superior Oblique Muscle

The superior oblique muscle is attached posteriorly behind the equator of the eye (see Fig. 4.2), elevates the back, and thereby depresses the front of the eyeball. This action increases as the eye is adducted; it is minimal when the eye is abducted. In the subsidiary action abduction and intorsion increase as the eye is abducted.

4.4.2 The Nerve and Blood Supply of the Superior Oblique Muscle

The fourth cranial nerve enters the muscle superiorly near its lateral border. The blood supply is from the superior muscular branches of the ophthalmic artery.

4.5 The Inferior Oblique Muscle

The inferior oblique muscle is the only muscle that takes its origin from the front of the orbit as a rounded tendon from a small depression on the orbital plate of the maxilla, a little behind the lower orbital margin and just lateral to the opening of the nasolacrimal duct. It passes backward laterally at an angle of 50° with the visual axis, roughly parallel to the tendon of the superior oblique muscle between the inferior rectus and the floor of the orbit. It then passes near the lateral rectus and is inserted by a very short tendon at the back and lateral portions of the globe, for the most part below the horizontal meridian. The line of insertion is oblique, 9.4 mm long, and is convex upward (see Fig. 4.8).

The inferior division of the third cranial nerve supplies the inferior oblique muscle; the nerve crosses above the posterior border of muscle about at its middle and enters the muscle on its upper surface. Its blood supply is from the infraorbital artery and the medial muscular branch of ophthalmic artery.

4.5.1 Principal Action of the Inferior Oblique Muscle

The principal action of this muscle is elevation, which increases as the eye is adducted. It is the only elevator in the adducted position of eyeball. Subsidiary actions are abduction and extorsion, which increase as the eye is abducted. The inferior oblique muscle and the superior rectus muscle act together to rotate the visual axis upward.

4.6 Levator Palpebrae Superioris

In this muscle's name the word palpebral means the lid, superioris denotes its presence in the upper lid, and the action is elevation of the upper lid.

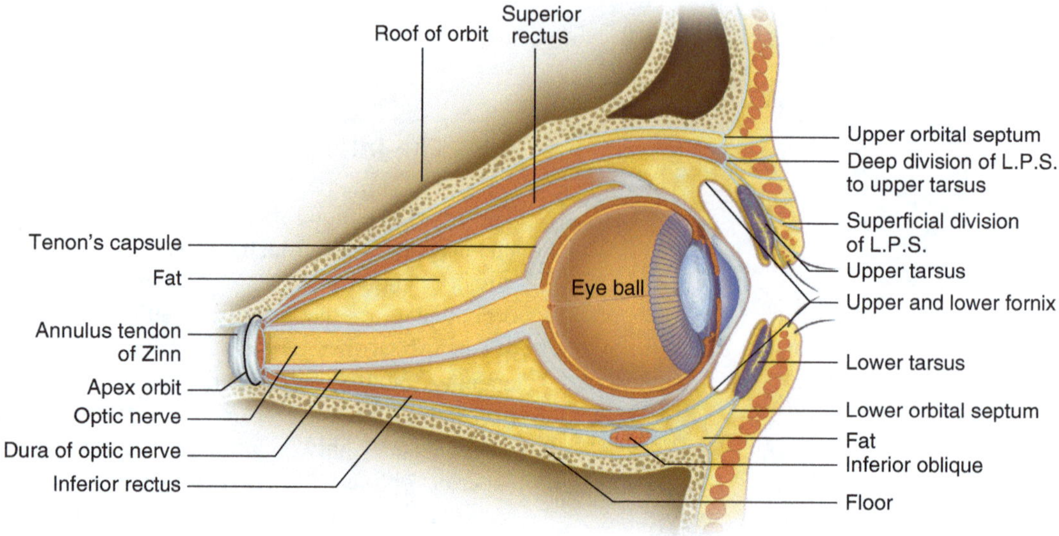

Fig. 4.9 Vertical anteroposterior section of the orbit. *LPS* levator palpebrae superioris

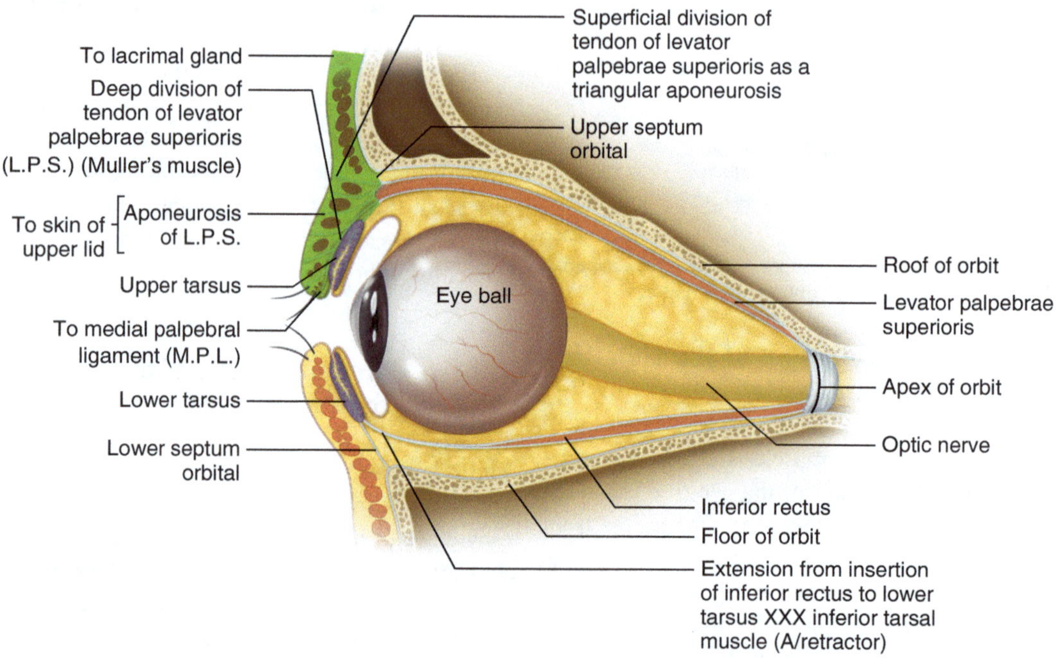

Fig. 4.10 The tendon of the levator muscle divides into a superficial and a deep part. The superficial part widens as a triangular aponeurosis (*green*), and the deep part goes to the *upper border* of the tarsus (Müller muscle). The base of the triangular aponeurosis gets inserted into the undersurface of the skin of the upper lid after passing through the orbicularis oculi. Its lateral end divides the lacrimal gland into an orbital and a palpebral part and its medial end goes to the medial palpebral ligament. Note how an extension from the insertion of the inferior rectus pierces the lower septum orbitale and gets inserted into the *lower border* of the inferior tarsus

The muscle arises from the undersurface of the lesser wing of the sphenoid above and in front of the optic foramen by means of a short tendon that is blended with the underlying origin of the superior rectus muscle (Figs. 4.9 and 4.10). The ribbonlike muscle belly passes forward below the roof of the orbit and on the superior rectus muscle (see Fig. 4.9) to about 1 cm, a few millimeters in front of the equator of the eye, where it ends in a membranous expansion called the aponeurosis of the levator palpebrae superioris. Actually the muscle divides into a deep muscular part attached to the upper border of the tarsus, forming the nonstriated superior palpebral muscle called the Müller muscle supplied by the cervical sympathetic nerve. The superficial part of the muscle, which has taken the shape of a triangular aponeurosis, molds itself on the globe of the eye. The change in direction takes place above the reflected tendon of the superior oblique muscle. The deeper muscular part is attached to the upper border of the upper tarsus, and the triangular superficial part (the aponeurosis) is clearly shown in Figs. 4.9 and 4.10 in a highly diagrammatic way. Both of these parts almost have to pass through the upper septum orbitale to reach the upper lid.

4.7 Insertions of the Levator Palpebrae Superioris

1. The deeper muscular part goes to the upper border of the upper tarsus.
2. The base of the triangular aponeurosis goes to the undersurface of the skin of the upper lid after passing through the subcutaneous palpebral part of the orbicularis oculi, along the superior palpebral furrow in the upper lid. The medial horn of the aponeurosis goes up to the medial palpebral ligament. The lateral horn of the triangular aponeurosis reaches the lacrimal gland to divide it (Fig. 4.11) into an orbital part and a palpebral part of the lacrimal gland, ultimately reaching the orbital tubercle and lateral palpebral ligament. The

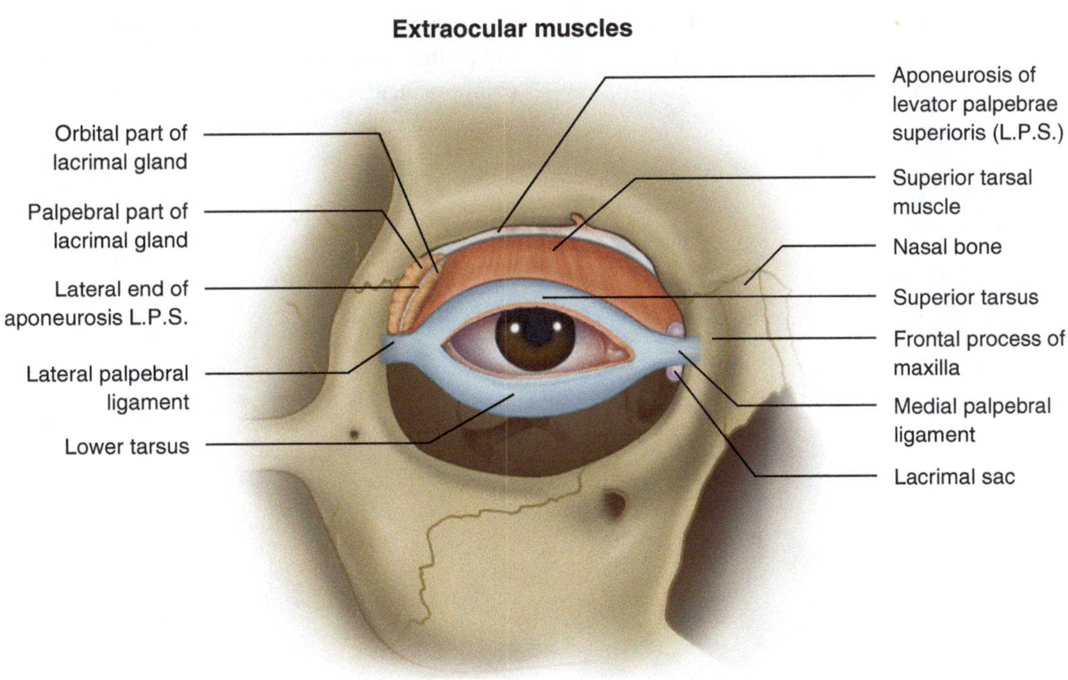

Fig. 4.11 Aponeurosis of the lateral end of the levator palpebrae superioris (LPS) dividing the lacrimal gland into an orbital and a palpebral part

sheath of LPS muscle attached to the sheath of the underlying superior rectus muscle as well as to the superior fornix.

4.8 Orbicularis Oculi (Sphincter of the Eyelids)

The orbicularis oculi forms an elliptical sheet surrounding the palpebral fissure, covers the lids, and spreads out for some distance to the forehead, temple, and cheek. It has two parts.

(1) Palpebral part: This is the central part of the muscle confined to the lids. It has pale fibers and may itself be divided into pretarsal and preseptal layers. The junction of the two, which is the thinnest portion of muscle, lies at the upper and lower lid furrows. Starting from the medial palpebral ligament and the nearby bone, the muscle passes across the lids in a series of half ellipses (Fig. 4.12) that meet outside the lateral canthus in the lateral palpebral raphae. This is made up of interdigitating muscle fibers strengthened by the septum orbitale. The fibers of the levator palpabrae superioris have to pass through this portion to reach the undersurface of the skin of the upper lid.

(2) The orbital portion has a curved origin from the medial side of the orbit, from the medial region of the margin of the upper orbital margin, medial to the supraorbital notch, from the maxillary process of the frontal bone, from the frontal process of the maxilla, from the medial palpebral ligament, and from the lower orbital margin medial to the infraorbital foramen. From this origin the peripheral fibers sweep across the orbital margin in a series of concentric loops, while the central fibers form almost complete rings.

A part of the orbicularis oculi is called the Horner muscle. It arises behind the lacrimal sac from the upper part of the posterior lacrimal crest and from the lacrimal fascia. It passes laterally and forward, divides into two slips that surround the canaliculi, and becomes continuous with the pretarsal fibers and the ciliary part of the orbicularis oculi (of Riolon muscle) in both lids.

4.9 Actions of the Orbicularis Oculi Muscle

The palpebral part controls involuntary blinking, holds the lower lid to the eyeball to prevent epiphora, and spreads a film of tears over the

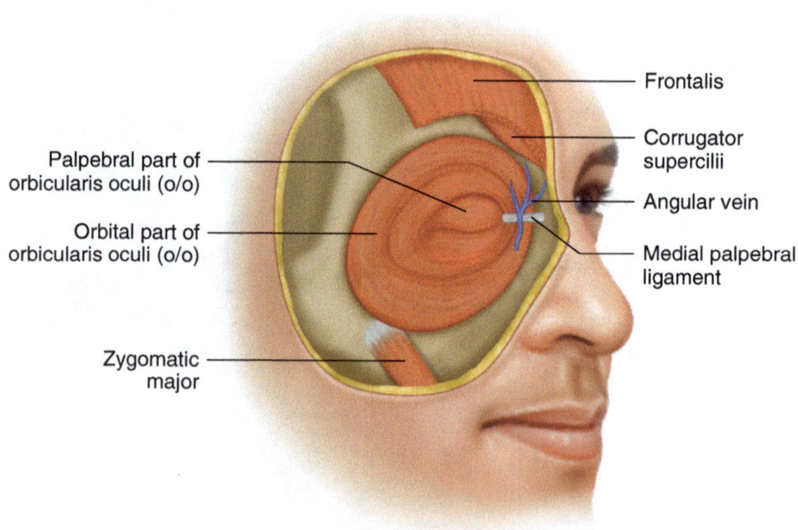

Fig. 4.12 The orbicularis oculi muscle (%) has two parts: (1) the palpebral part used in blinking and (2) the orbital part used in forced closure of the lids

eyeball to moisten it. It is opposed by the levator palpabrae superioris.

The orbital part acts to close the eyelids very tightly.

The Horner muscle dilates the lacrimal sac.

The pumping action of the orbicularis oculi is clearly seen in blinking movements that remove excess tears.

The orbicularis oculi is supplied by the facial nerve: the upper part by the temporal and upper zygomatic and the lower part by the lower zygomatic branches. One part of the muscle may be paralyzed without the other.

4.10 The Intraocular Muscles

There are three involuntary, unstriated muscles inside the eye. Two are present in the iris (see Fig. 2.8, Chap. 2, p. 18). The first is the sphincter pupillae, seen as a circularly arranged ringlike muscle around the pupil. It is supplied by the parasympathetic nervous system via a branch of the third cranial nerve. The branch arises from the nerve to the inferior oblique muscle as a root of the ciliary ganglion, where a relay occurs and a new neuron takes over to continue as short posterior ciliary nerves. The second consists of dilator fibers (see Fig. 2.8, Chap. 2, p. 18) that are accompanied by pigment attached to the peripheral border of the sphincter at the stroma of the iris. The third involuntary muscle is found inside the ciliary body and is called the ciliary muscle (see Fig. 2.12, Chap. 2, p. 21). It has the form of a right-angled triangle; the right angle is internal and faces the ciliary processes. The posterior angle is acute and points to the choroid, and the hypotenuse runs parallel to the sclera. This muscle has three types of fibers externally: outermost is the longitudinal, the intermediate oblique or radial, and the innermost circular or sphincteric. The longitudinal fibers have their origin in the scleral spur and the adjacent trabeculae. They can be traced posteriorly to the suprachoroid lamina. They open the trabecular spaces and the canal of Schlemm. The circular fibers occupy the most internal part of the ciliary body. They form a ring whose contraction reduces the diameter of the ciliary ring, which in turn releases the tension of the zonule of the lens. The zonule sets free the tension of the capsule of lens, thus producing an increased curvature of the lens at the anterior pole. This increased curvature converges the incident divergent rays coming from nearby objects. This process of self-focusing on a near object is called accommodation. Accommodation becomes weaker with aging because of the development of hard lens matter that cannot be molded easily. Thus the longitudinal fibers increase the outflow of the aqueous to adjust intraocular pressure while the circular fibers help in accommodation in patients under 40 years in whom lens matter is moldable (soft).

Suggested Reading

Parsons' Diseases of the Eye, edn 16. Revised by Miller SJH. London: Churchill Livingstone; 1979. pp. 1–14; 545–60

Riordon-Eva P, Cunningham E (2011) Vaughan and Asbury's general ophthalmology, edn 18. McGraw Hill, New York, pp 1–26

Trevor-Roper PD (1974) The eye and its disorders. Blackwell Scientific Publications, London, pp 26–44, 323–34

Wolff's E (1976) Anatomy of the eye and orbit, edn 7. Revised by Warwick R. London: HK Lewis; 1976; pp. 1–29; 30–180D

Anatomy of the Eyelids

5

Keywords

Skin–muscle layer · Tarsoconjunctival layer · Gray line · Posterior sharp border of lid margin · Anterior round border of lid margin · Hair follicle gland of Zeis · Moll's gland · Meibomian gland · Three-layered tear film · Two divisions of the tendon of the levator palpebrae superioris · Triangular aponeurosis · Lateral cornua · Medial and lateral palpebral ligament · Superior tarsus · Inferior tarsus · Septum orbitale

The eyelids are made up of two parts: the frontal skin–muscle layer is the soft layer, and the back hard segment is the tarsoconjunctival layer. They can easily be separated with a knife when entering along the gray line at the lid margin (Fig. 5.1). The skin of the lids is the thinnest in the body, almost transparent. It has no subcutaneous fat. It has a striated subcutaneous muscle in the upper and lower lid called the palpebral part of the orbicularis oculi through which the superficial division of the tendon of levator passes. The tendon is inserted into the undersurface of the skin of the upper lid along a line parallel to the superior palpebral furrow. The skin can easily be lifted from the subcutaneous muscle. The subcutaneous muscle is responsible for the involuntary rhythmic closure of the lid called blinking.

The lid margin is about 2 mm thick, horizontally placed, and has a sharp posterior border and a round anterior border. The round anterior border has hair follicles of the eyelashes, into which the sebaceous gland of Zeis opens (Fig. 5.2). When the follicle is blocked, the resident bacteria multiply and cause a suppurative inflammation of the Zeis gland called a stye, which is always related to an eyelash. Just anterior to the sharp posterior border of the lid margin is the opening of the special glands present in the substance of tarsus, called the meibomian glands (Fig. 5.3). The sharp posterior border of the lid margin holds a tear streak along its length. This streak is covered with an oily layer from the meibomian gland to prevent the spilling of tears. This sharp posterior border remains in contact with the eyeball in all positions of gaze of the eye.

The hard part of the lid is called the tarsus; it gives shape to the lid. It has an anterior and a posterior surface. Its upper border is connected to the superior orbital margin by a sheet of fascia called the septum orbitale in the upper lid. In the

M.W. Ansari and A. Nadeem, *Atlas of Ocular Anatomy*, DOI 10.1007/978-3-319-42781-2_5

lower lid, it is the lower border of the tarsus, which is connected to the inferior orbital margin by the lower septum orbitale. The septum orbitale is not a rigid structure but a floating membrane that molds itself to the shape of the lid and moves with all the movements of the lid. In the upper lid, the upper septum orbitale is pierced by the superficial division of the tendon of the levator palpebrae superioris (LPS); in the lower lid, the lower septum is pierced by a prolongation of the inferior rectus muscle, called the inferior tarsal (Figs. 5.4 and 5.5) muscle. Just anterior to the opening of the meibomian gland is a gray line along which the lid can be vertically split into its anterior soft skin–muscle layer and the hard posterior tarso conjunctival layer by a knife. The innermost layer of the palpebral conjunctiva is adherent to the tarsus. The integrity of the three-layered tear film is maintained by the swab-like blinking movement of the upper lid, which spreads the tears on the front of the cornea and conjunctiva. Its outermost oily layer comes from the meibomian gland, which prevents any spilling of tears on the cheek. The middle watery layer comes from the main and accessory lacrimal glands. The innermost mucoid layer of tear film comes from mucus-secreting conjunctival glands. It keeps the cornea and conjunctiva moist and prevents their dehydration.

The eyelids act as shutters protecting the eye from injury or excessive light. They thus aid the pupil in controlling the amount of light that reaches the retina. The visual cortex is at rest when the lids are shut. Regarding tears, lids have a dual function. Blinking spreads the tear film over the cornea like a swab, protecting it from dehydration. Blinking is a rhythmic reflex, fired off by evaporation. Blinking empties the conjunctival sac by its pumping effect on the lacrimal sac. The upper lid extends above to the eyebrow, which separates it from the forehead; the lower lid passes without a line of demarcation into the skin of the cheek. The eyebrows are folds of thickened skin covered with hair, which is supported by underlying muscle fibers. The glabella is the hairless prominence between the eyebrows. As mentioned in Chap. 1, in the primary position of the eyeball the lower margin of the upper lid just touches the upper limbus, and the upper margin of the lower lid just touches the lower limbus; and the size of this palpebral fissure is normally equal on the two sides. In irritations caused by inflammation of the anterior segment of the eye (conjunctivitis, keratitis, scleritis, or anterior uveitis), a reflex spasm of that part of the orbicularis oculi responsible for blinking causes a narrow palpebral fissure on that side. During closure of the lids the eyeball goes up, which is called the Bell phenomenon. This explains why that part of the cornea just below the center will suffer from degeneration or exposure keratitis. The lateral and medial ends of the lids join each other, forming the outer and inner canthus. The outer canthus is about 2 mm higher than the inner one. The outer canthus lies on the globe, and the medial canthus is separated from the globe by a little bay—the tear lake. In this, the caruncle can be seen on the medial side of a crescentic fold of conjunctiva called the plica semiluranis.

5.1 Structure of the Eyelids

The following are the layers of the eyelid from anterior to posterior:

1. The skin
2. A layer of subcutaneous areolar tissue
3. A layer of striated muscle (the orbicularis oculi)
4. The submuscular areolar tissue
5. A hard layer that gives shape to the lids called the tarsus
6. A layer of nonstriated muscle
7. The mucous membrane–palpebral conjunctiva

1. The skin of the lids is almost the thinnest in the body; it is almost transparent so that it forms folds and is easily wrinkled. It is very elastic and recovers rapidly after being distended by fluids (bruises). When eye is open, a sulcus is seen along the upper border of tarsus called the superior palpebral furrow; it is produced by the pull of the tendon of the LPS. The corresponding furrow in

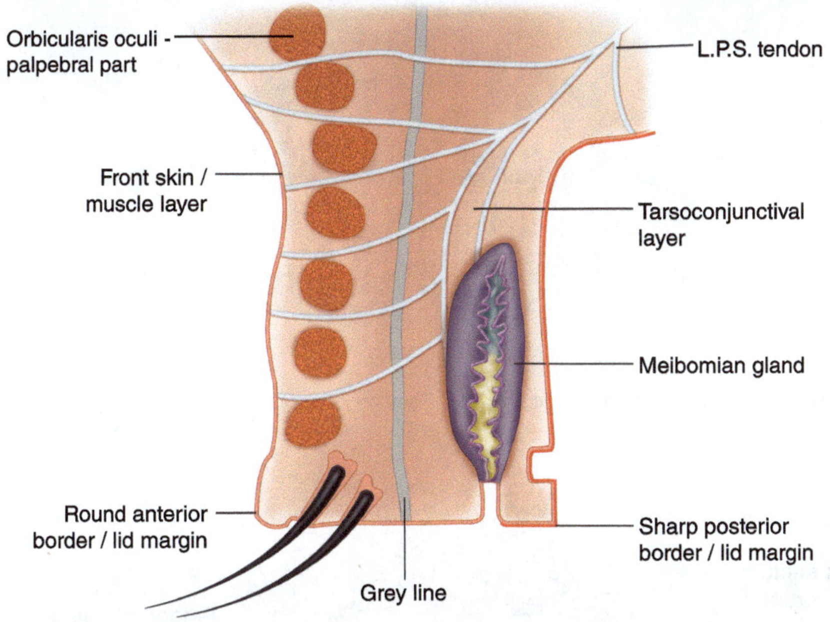

Fig. 5.1 Skin: muscle and tarso conjunctival layers of the lid along the *gray line*

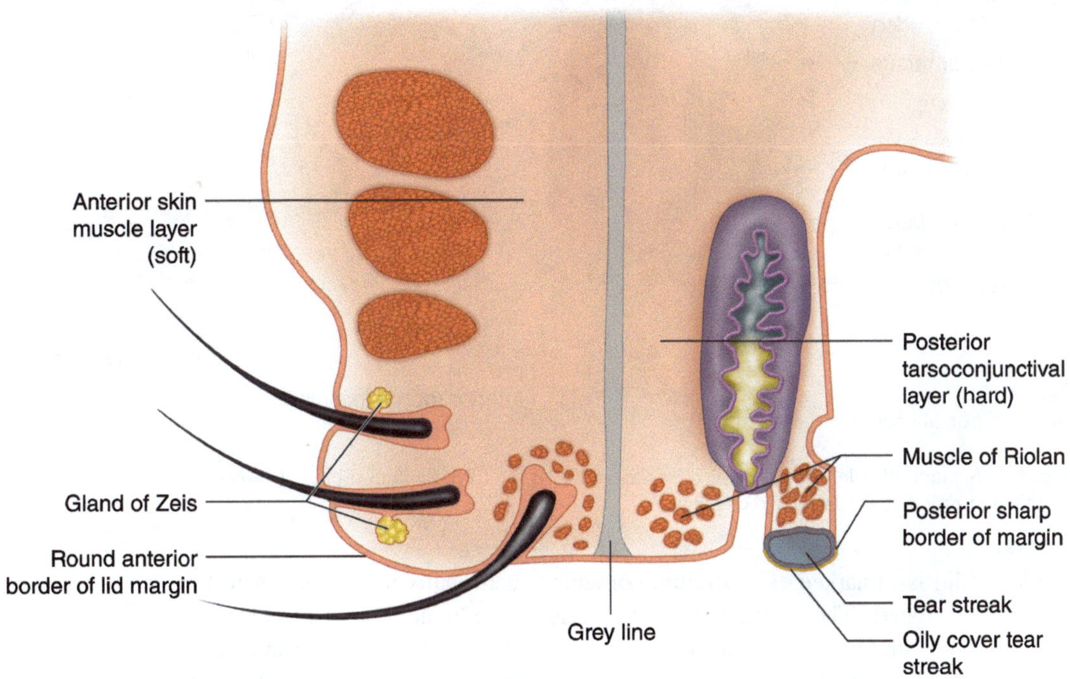

Fig. 5.2 Skin: muscle and tarso conjunctival layers of the lid. Note the sharp posterior border of the lid margin with tear drops covered by an oily secretion from the meibomian glands. Note also the round anterior border of the lid margin with the hair follicles of the eyelashes

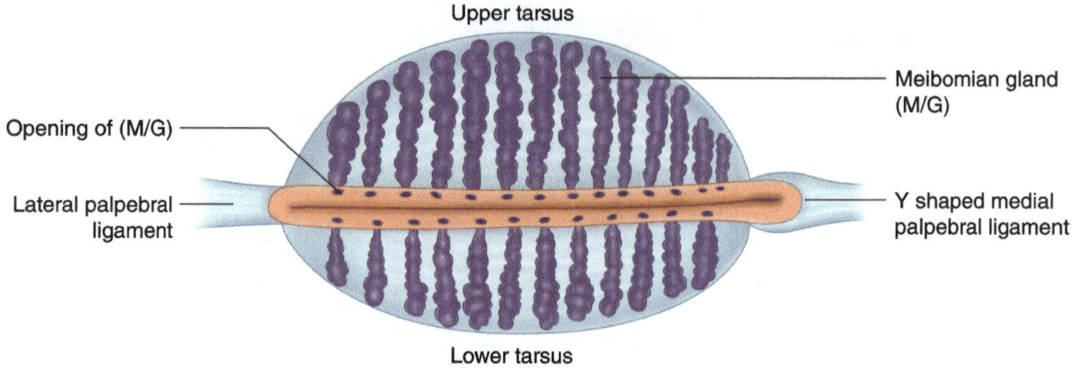

Fig. 5.3 Meibomian glands seen on everting the lids. They are at a right angle to the margin of the lid

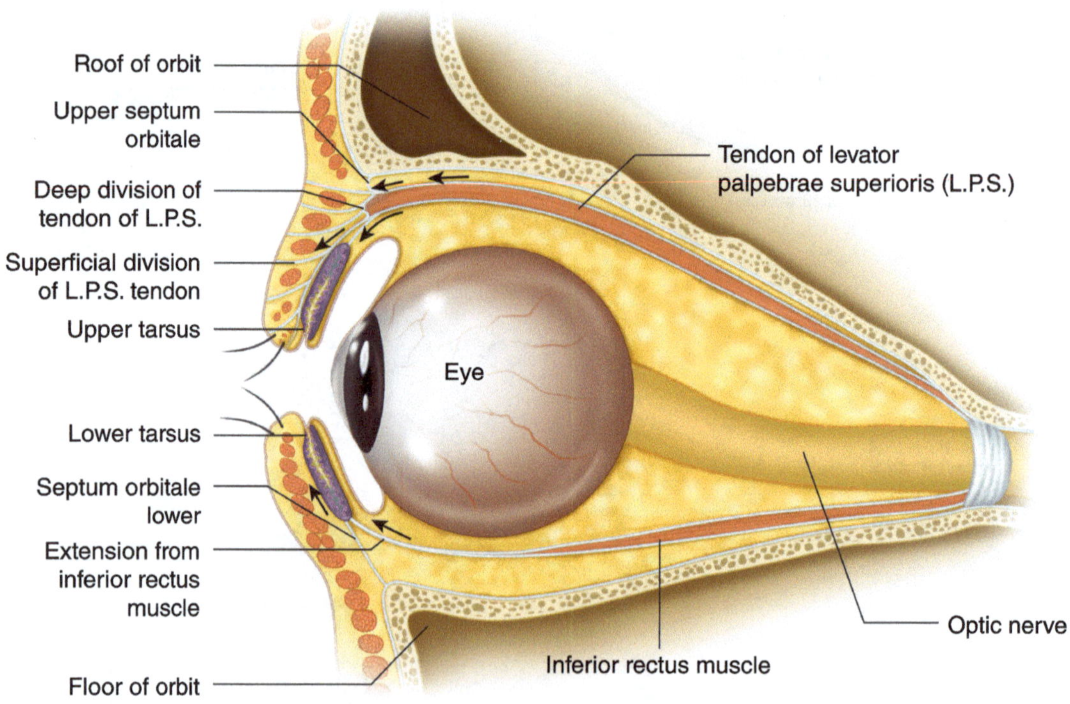

Fig. 5.4 Septum orbitale (*upper*) pierced by the tendon of the LPS. Septum orbitale (*lower*) pierced by an extension from the insertion of the inferior rectus muscle

the lower lid is illmarked. The stratum corneum is well developed. The stratum granulosum is present. The stratum mucosum has three or four layers of cells. Then comes the stratum germinativum, which is resting on a basement membrane. The nasal and temporal portions of the skin of the lid differ; the nasal skin is smoother, shinier, and greasier and has fewer hairs.

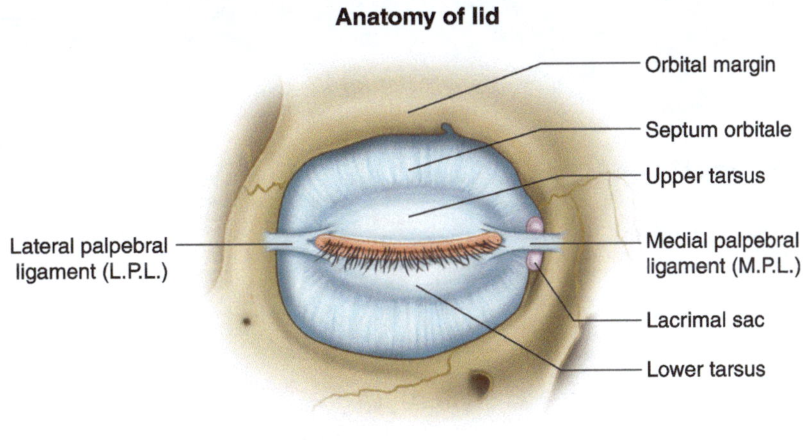

Fig. 5.5 Septum orbitale and two tarsi

2. The subcutaneous areolar layer consists of loose connective tissue without any fat so that the skin can easily be lifted by effusion with blood (ecchymosis).
3. The third is the layer of the subcutaneous palpebral part of the orbicularis oculi (see Figs. 5.1 and 5.2), which is used in the involuntary rhythmic closure of the upper lid, called blinking. The muscle fibers are arranged concentrically around the palpebral opening. It is the part of the muscle through which the superficial division of the tendon of the LPS has to pass through before being inserted on the undersurface (Fig. 5.6) of the skin of the upper lid along a line demarcated by the superior palpebral furrow. The part of the orbicularis oculi that lies next to and occupies the whole thickness of the lid margin is called the muscle of Riolan. It is obliquely traversed by the follicles of the lashes, the glands of Moll, and the meibomian glands (see Figs. 5.6).
4. The submuscular areolar tissue lies between the orbicularis oculi and the tarsal plate and communicates above with the subaponeurotic stratum of the scalp, the so-called dangerous area of the surgeons. Hence pus or blood can make its way into the upper lid from the dangerous area. It is through this plane, reached by entering the knife at the gray line, that the lid may be split into anterior and posterior portions the anterior soft portion (skin and muscle layer) and the posterior hard portion; the latter is the tarso conjunctival layer. The main nerves to the lids also lie in this areolar tissue for infiltration anesthesia. The local anesthetic has to be injected deep into the orbicularis. In the lower lid, this tissue lies in a single small space, the preseptal space in front of the septum orbitale. In the upper lid, the space in which this tissue lies is divided by the LPS into pretarsal and preseptal spaces. The pretarsal space is small and contains the peripheral arterial arcade (see Fig. 5.6); the preseptal space is triangular on a vertical section. In front of it is the orbicularis and behind it is the septum and those tendinous fibers of the LPS that pierce the orbicularis (see Fig. 5.6).
5. The fibrous layer gives a shape to the lids. It has a central, very thick, and hard part—the tarsal plates—and a thinner peripheral part called the palpebral fascia or the septum orbitale. There is one tarsal plate for each lid, forming the skeleton of lids and giving them their shape and firmness. The crescent-shaped upper tarsus is much larger: 11 mm in height at its middle. The lower plate is only 5 mm high. Each tarsus is about 29 mm long and 1 mm thick, has an anterior and a posterior surface, and both a free and an attached border. The medial end is separated from the orbicularis by loose areolar tissue so that the muscle moves freely on the tarsus. The posterior surface, which is concave and lined

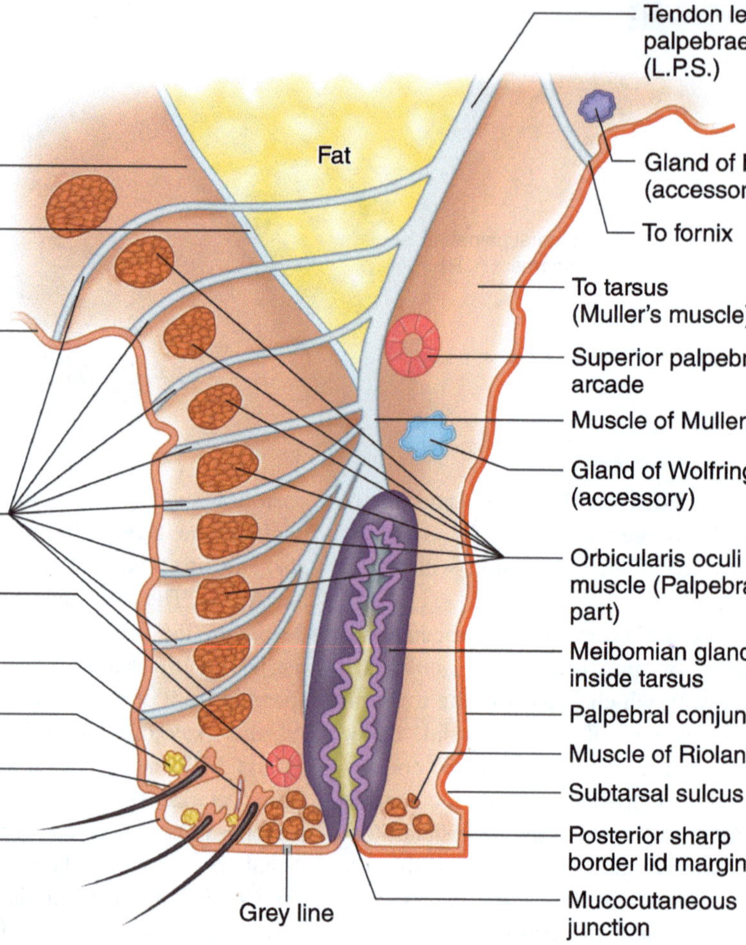

Fig. 5.6 Highly diagrammatic vertical section of the upper lid showing the sharp posterior border of the lid margin, the round anterior border of the lid margin with hair follicles and the gland of Zeis, the Moll's gland, and the meibomian glands. Note the multiple insertions of the LPS tendon

with adherent palpebral conjunctiva, molds itself on the eyeball. The free border, forming the thick, almost horizontal margin of the lid is about 2 mm broad. The lid margin has a sharp posterior border that remains in contact with the eyeball in all positions of gaze and has a horizontal streak of tears all along its length, covered by an outer oily secretion coming from the meibomian glands. This oily covering of tear streak along the sharp posterior border of the lid margin prevents spilling of tears. The attached border of the tarsus is thin and gradually runs into the septum orbitale. The lateral and medial extremities of the tarsal plates are attached to the orbital margin by strong fibrous structures called the lateral palpebral and medial palpebral ligaments. The tarsal plates contain sebaceous glands, which run from their attached margin to their free margins. They are arranged vertically, parallel with each other, at a right angle to the lid margin; there are about 25 in the upper tarsus and 20 in the lower tarsus. Each consists of a central canal; numerous

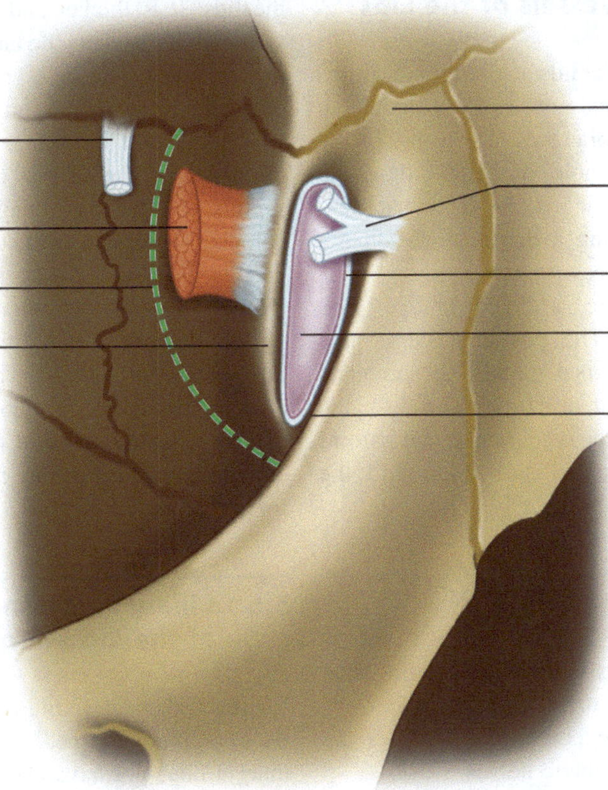

Fig. 5.7 Lacrimal sac in the lacrimal fossa between the anterior and posterior lacrimal crests below the inner canthus

rounded appendages that secrete sebum open into their sides. These glands are called the meibomian glands. The small openings of the canal can be seen on the free margins of the lid just in front of its sharp posterior border. These glands can easily be seen through the palpebral conjunctiva as yellow streaks when you evert the lids.

6. The sixth layer is composed of nonstriated muscle fibers. In the upper lid, the deep division of the tendon of the LPS lies just deep to the septum orbitale and gets attached to the upper border of the upper tarsus and is called the muscle of Müller. In the lower lid, a prolongation of the inferior rectus muscle gets attached to the lower border of the lower tarsus after piercing the lower septum orbitale this is called the inferior tarsal muscle. The Müller muscle and the inferior tarsal muscle are supplied by the cervical sympathetic nerve, which explains ptosis, drooping of the upper lid in Horner syndrome.

7. The palpebral conjunctiva is firmly adherent to the inner surface of both tarsi.

5.2 Glands of the Lids

(1) The meibomian glands are described above.
(2) The glands of Moll are 1–2 mm long and consist of simple spiral, modified sweat glands placed obliquely in contact with and parallel to the bulbs of the cilia. The duct passes through the dermis and epidermis and may terminate separately between two lashes or may open into the duct of a sebaceous gland of Zeis.
(3) The sebaceous glands of Zeis are modified sebaceous glands attached directly to the follicles of the eyelashes. Their suppurative inflammation gives rise to a painful yellow nodule at the root of an eyelash called a stye.

5.3 The Blood Vessels of the Lids

The blood vessels of the lids are derived from the ophthalmic and lacrimal arteries by their medial and lateral palpebral branches. The veins of the lids are larger and more numerous than the arteries and are arranged as pretarsal and post-tarsal sets, forming a dense plexus in the upper and lower fornices.

The lymphatics are arranged in pre- and post-tarsal plexuses. The post-tarsal drain the conjunctiva and the tarsal glands; the pretarsal drain the skin and skin structures. Those for the lateral side drain into the preauricular and parotid nodes; those from medial side drain into the submandibular lymph nodes.

5.4 Nerves of the Lids

The muscle orbicularis is supplied by the facial nerves; the LPS by the upper division of the third cranial nerve, and the nonstriated muscles are supplied by the sympathetic nerves.

Sensory: The upper lid is supplied by the supraorbital nerve. On the medial side, the supra- and infratrochlear and, on the lateral side, the lacrimal branches of the ophthalmic division of trigeminal assist. The lower lid receives its supply from the infraorbital nerve.

5.5 Septum Orbitale

As mentioned in Chap. 1, the septum orbitale is attached above to the orbital margin along a thickening called the arcus marginale; centrally it is in continuation with the tarsal plates. In the upper part it is pierced by the tendon of the LPS (its superficial division), and in the lower lid it is pierced by a prolongation from the inferior rectus muscle that is inserted at the lower border of the lower tarsus. The septum orbitale is not a fixed and tight structure but actually is a floating membrane and takes part in all the movements of the lids. It is thicker and stronger on its lateral side than its medial side and in the upper lid than in the lower. The weak portions of the septum determine the site of herniation of the orbital fat, which lies just deep into it. These hernias are often seen in the elderly. On the lateral side, the septum is superficial, lying anterior to the lateral palpebral ligament, while on the medial side it is deep, lying behind the lacrimal part of orbicularis (Horner muscle, see Fig. 5.7 on p. 59). The septum separates the orbital contents from the lids so that inflammation anterior to the septum does not reach the orbit too quickly. If this occurs orbital inflammation will take longer to present clinically because it may remain restricted there for sometime.

5.6 Ciliary and Nonciliary Parts of the Lid Margin

The eyelashes are found in the lateral part of the lid margin. The medial-most part of the lid margin does not have eyelashes. At the junction of the ciliary and nonciliary parts of the lid margin is a small papilla with a hole in it. This hole is called the upper and lower punctum and normally dips into the microscopic lake of tears inside the inner canthus. Each punctum opens into a vertical part of a very thin canal in the nonciliary part of the lid margin, which is continuous with a horizontal part of the canaliculus. The canaliculus separately opens into the lateral wall of the fundus of the lacrimal sac, lodged in a fossa in the medial wall of the orbit (Fig. 5.7). This lacrimal sac opens down below into a nasolacrimal duct whose lower end opens into the inferior meatus of the nose. The puncta, canaliculi, lacrimal sac, and nasolacrimal duct together are called the drainage part of the lacrimal apparatus. The secretory part of the lacrimal system consists of the lacrimal gland proper in the lateral part of the roof and two accessory lacrimal glands called the glands of Krause and Wolfring near the superior fornix.

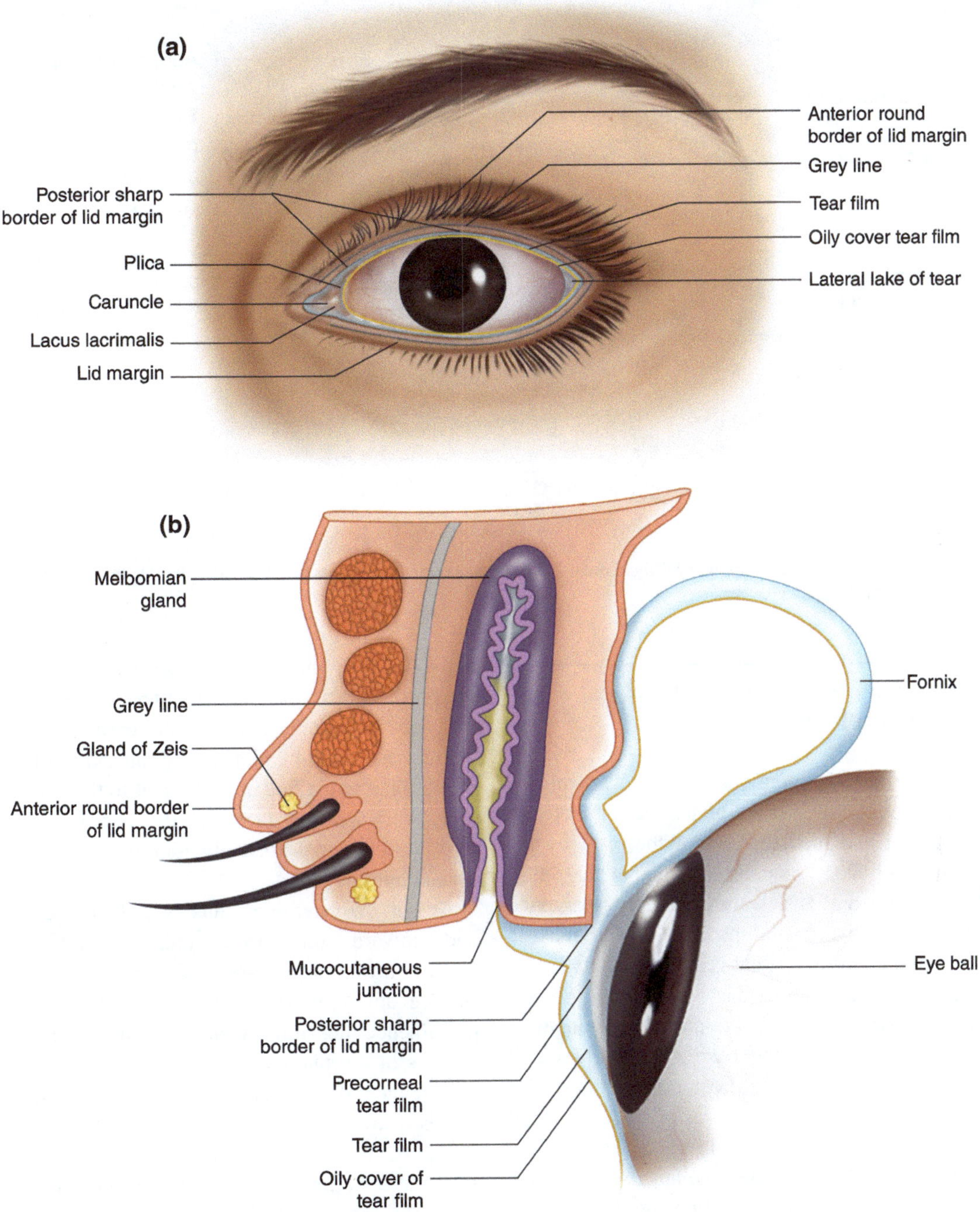

Fig. 5.8 **a** and **b** Detailed anatomy of the lid margins to show the tear bead just anterior to the sharp posterior border of the lid margin covered by an oily secretion from the meibomian gland to prevent spilling of tears

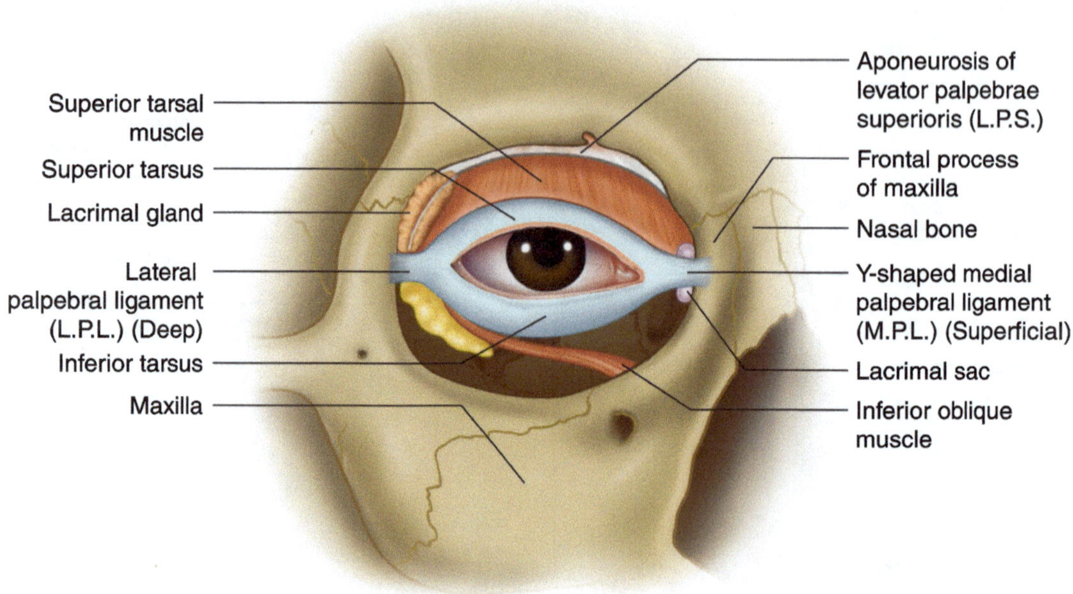

Fig. 5.9 Medial and lateral palpebral ligaments. Note the lower end of the lacrimal gland touching the back of the lateral palpebral ligament

5.7 The Medial and Lateral Palpebral Ligaments

We have already mentioned that the lateral and medial ends of both tarsal plates are connected to the lateral and medial walls of the orbit by the medial and lateral palpebral ligaments (Fig. 5.8). The medial one is a somewhat triangular band attached to the frontal process of the maxilla from the anterior lacrimal crest to near its suture with the nasal bone. It has a lower free border, while above it is adherent to and continuous with the periosteum. At the base of the triangle it divides into anterior and posterior portions. The posterior portion is continuous with the lacrimal fascia and thus helps to place a roof over the upper part of lacrimal sac. The anterior portion is continued at the medial canthus and divided into two bands that pass across the lacrimal fossa (but not in contact with the sac). They attach to the medial extremities of the tarsal plates. These bands make an angle laterally with the lacrimal fascia. They form the letter Y placed on its side with the main ligament. The two branches that correspond to the lacrimal portions of the lid margins, and in fact are tubular, contain the lacrimal canaliculi, enclose the caruncle, and delimit the medial canthus. The anterior surface of the ligament is free and adherent to the skin. It looks forward and laterally; the two branches look forward and medially and thus make an obtuse angle.

When the lateral canthus is pulled laterally and upward, the medial palpebral ligament (MPL) forms a well-marked prominence for surface anatomy. It should be noted that this prominence lies almost entirely on the frontal process of the maxilla.

The lateral palpebral ligament (Fig. 5.9) is attached to Whitnall's tubercle on the zygomatic bone, 11 mm below the front of the zygomatic suture. It is 7 × 2.5 mm, lies deeper, and does not form a prominence like the MPL. Its front surface is fused with the pre-ciliary fibers of the orbicularis. Its posterior surface is in front of the lateral check ligament and separated from it by a lobule of the lacrimal gland (see Fig. 5.9). Its upper border is united with the expansion of the LPS, its lower border with the expansion from the inferior oblique and inferior rectus muscles.

Suggested Reading

Parsons' Diseases of the Eye, edn 16. Revised by Miller SJH. London: Churchill Livingstone; 1979. pp. 1–14; 545–60

Riordon-Eva P, Cunningham E (2011) Vaughan and Asbury's general ophthalmology, edn 18. McGraw Hill, New York, pp 1–26

Trevor-Roper PD (1974) The eye and its disorders. Blackwell Scientific Publications, London, pp 26–44, 323–34

Wolff's E (1976) Anatomy of the eye and orbit, edn 7. Revised by Warwick R. London: HK Lewis; 1976; pp. 1–29; 30–180D

Transparent Structures of the Eyeball Cornea, Lens, and Vitreous

6

Keywords

Transparent dome · Cornea · Bowman's membrane · Tear film · Descemet's membrane · Dehydrating pump · Endothelium · Superficial keratitis · Deep keratitis · Capsule of lens · Differential thickness · Zonule · Accommodation · Smoothing action of tear film for microscopic irregularities of corneal epithelium · Optimum width of angle of anterior chamber · Senile cataract · Nutrition of cornea and lens by normal aqueous

The eyeball resembles a ping-pong ball with a transparent dome in the front called the cornea. The cornea constitutes one-sixth of the eyeball; the posterior five-sixths are formed by the scleral shell. The cornea is thin in the center and thick at the periphery. Superiorly and inferiorly it is encroached upon by the sclera in a crescentic way, thus reducing the frontal vertical diameter. It is circular in outline; the periphery merges with the sclera, and the junction is called the limbus. Its front surface is covered by a three-layered tear film that keeps it moist and smooths the microscopic irregularity of its outermost layer: the corneal epithelium. That smooth surface is necessary for proper refraction at the cornea. It has five layers: the epithelium, Bowman's membrane, the stroma, Descemet's membrane, and the endothelium. The cornea is transparent because of its avascularity and the regular arrangement of its cellular components. This transparency is maintained by a dehydrating pumping mechanism chiefly located in its endothelium and partly in the epithelium that allows only an optimum quantity of water in its stroma. Any condition that interferes with this dehydrating pump, such as a sudden gross rise of intraocular pressure as occurs in acute congestive glaucoma or deep keratitis, will allow more than the optimum amount of water inside, making it cloudy. Superficial keratitis anterior to the Bowman's membrane heals without scarring, and therefore no opacity results. However, any keratitis deeper than Bowmanh's membrane will heal by scarring, leading to corneal opacity. This will reduce vision if it is in the pupillary line.

The nutrition of the cornea comes directly from the air (oxygen), the tear film on its front, and the aqueous at its back. An abnormal aqueous (as in anterior uveitis) will give rise to precipitation of inflammatory cells on the back of the cornea called keratic precipitates. The corneal epithelium is continuous with the epithelium of the bulbar

M.W. Ansari and A. Nadeem, *Atlas of Ocular Anatomy*, DOI 10.1007/978-3-319-42781-2_6

conjunctiva and therefore it can be secondarily affected by conjunctival diseases like gonococcal and pseudomonal conjunctivitis. The cornea forms the anterior boundary of the anterior chamber, whose periphery is called the angle of the anterior chamber. The angle contains the microscopic outlet channels of the aqueous. The size of the angle of the anterior chamber must have an optimum width for an adequate outflow of aqueous to occur. A narrow angle can reduce aqueous outflow and may cause a rise in intraocular pressure, which if prolonged can damage the optic nerve. The rich sensory supply of cornea causes the pain in keratitis (inflammation of the cornea).

The lens lies in the coronal plane just behind the iris. It is held in position by the zonule, which is connected to the equator of the lens and the ciliary processes. The function of the lens is to further bend the refracted rays of light to form a clear image on the fovea. The more important function of the lens is accommodation, the self-focusing device of the eye to focus objects clearly situated between normal reading distances of 14 inches and infinity in patients below 40 years. In older patients, the lack of moldability of the lens material does not allow the increased curvature of the lens at its anterior pole. This creates problems in reading for which patients have to use a converging lens otherwise called reading glasses.

The transparency of the lens is maintained by its capsule, which allows only an optimum quantity of water inside the lens. Since the cause of a senile cataract is still not known, its treatment is only to remove the opaque lens and replace it in the same sitting with a refined plastic intraocular implant whose power can be calculated beforehand.

6.1 Medical Optics

Before talking about the transparent structures of eye it is useful to provide some basic concepts of medical optics. We know that light energy travels in a straight line called light rays. Whenever light rays travel from one transparent medium to another, they change their path. This change in their path is called refraction, which occurs according to some laws of higher mathematics. What we need to know is that incident light rays that come from 20 feet or more are supposed to be parallel. This explains why visual activity is tested at that distance, so that the patient need not use his or her self-focusing device called accommodation. Light rays coming from a distance less than 20 feet away are supposed to be divergent, and to focus them clearly on the retina, one has to use the self-focusing device of the eye called accommodation by increasing the curvature of the lens at its anterior pole (see Chap. 2). When all incident rays of light coming from a point at the object actually meet (or appear to meet) at another point, the second point is called the image of the first point. Transparent optical devices with suitable curvature (convex or concave) are used to achieve that, i.e., making the refracted rays meet at a suitable point. From a refraction point of view, a normal eye is that which can focus incoming parallel rays on its fovea without using accommodation. The human eye has a self-focusing device to focus objects from a reading distance of 14 inches to infinity. This device is called accommodation and is achieved by involuntary action of a muscle ciliary muscle found in the ciliary body (see Chap. 2).

6.2 Transparent Structures: The Cornea

The first transparent structure of the eye is the cornea, the transparent dome on its front, which forms the frontal one-sixth of the eyeball; the posterior five-sixths are formed by the scleral shell (see Chap. 1). The vertical diameter on its front is slightly less than the horizontal one because of a crescentic encroachment of the sclera above and below (Fig. 6.1a), while on its back both of the diameters are equal (Fig. 6.1b). Its periphery is thicker than the center (Fig. 6.1c). It is covered on its front surface by a three-layered

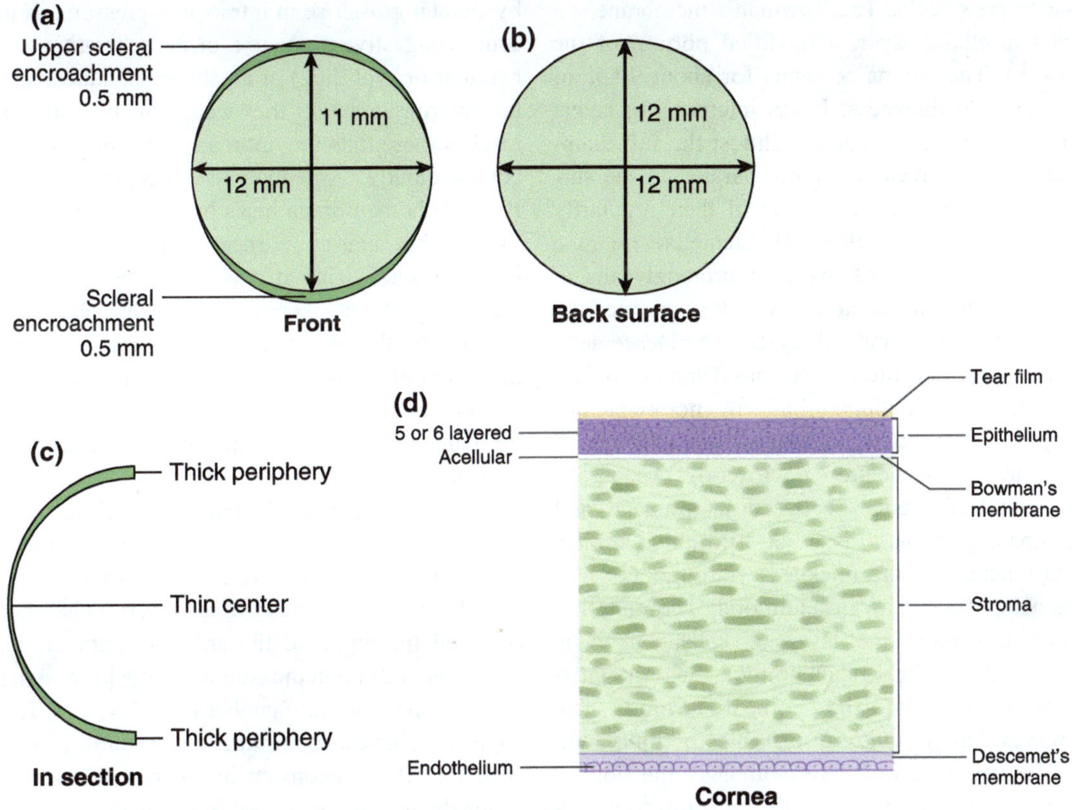

Fig. 6.1 **a** Cornea front view. **b** Cornea back surface. **c** Cornea in section, thinner in center, thicker at periphery. **d** Microscopic transverse section of cornea—five layers with tear film

tear film (see Chap. 5; Fig. 6.1d). The tear film serves many purposes. The outermost oily layer comes from the meibomian glands and regulates evaporation from its surface, keeping it at an optimum level. This evaporation takes heat from the surface, thus reducing the temperature of the conjunctival sac, which prevents multiplication of resident bacterial flora of the conjunctival sac. The lysozyme of the tear also does the same. The middle watery layer of tear film comes from the main lacrimal and accessory lacrimal glands. Normally, the main lacrimal gland is not used. The secretions from the accessory lacrimal glands of Krause and Wolfring are enough to keep both the conjunctiva and the cornea moist, and there is no need for drainage to the nose via the puncta, canaliculi, lacrimal sac, and nasolacrimal duct. The oxygen from the tears is used for corneal metabolism. The mucous layer, the innermost layer of tear film, comes from mucous-secreting glands of the conjunctiva and makes the surface smooth, removing the microscopic irregularities of the cells of the corneal epithelium. The involuntary act of the blinking of the lids spreads the tear film in front of the eyeball like a swab preventing its dehydration.

6.3 Microscopic Structure of the Cornea

From anterior to posterior the cornea has five definite layers: (1) the epithelium, (2) Bowman's membrane, (3) the stroma, (4) Descemet's membrane, and (5) the endothelium (Fig. 6.1). The outermost layer of the corneal epithelium is peripherally continuous with the epithelium of the bulbar conjunctiva at the limbus. It has five or six layers of cells. The Bowman's membrane is a clear acellular layer, a modified portion of the stroma. The stroma accounts for about 90 % of the corneal thickness. It has intertwining layers of collagen fibrils that run almost the full diameter of the cornea. They run parallel to the surface of cornea and by virtue of their regularity they are optically clear. The lamellae lie in a ground substance of hydrated proteoglycans in association with the keratocytes that produce the collagen and ground substance. The Descemet's membrane, constituting the basal lamina of the corneal endothelium, has a homogeneous appearance on light microscopy but a laminated appearance on electron microscopy because of structural differences between its prenatal and postnatal portions. It is about 3 μm thick at birth but increases in thickness throughout life, reaching 10–12 μm in adulthood. The endothelium has only one layer of cells, but it is responsible for maintaining the essential detergescence of the corneal stroma. The endothelium is quite susceptible to injury and undergoes a loss of cells with age, the normal density reducing from 23,000 cells/mm^2 at birth to 2000 cells/mm^2 in old age. Endothelial repair is limited to enlargement of existing cells with little capacity for cell division. Failure of endothelial function leads to corneal edema.

6.4 Nutrition of the Cornea

The sources of nutrition of the cornea are the vessels of the limbus, the aqueous, and tears. The superficial cornea also gets most of its oxygen from the atmosphere directly. The sensory nerves of the cornea are supplied by the ophthalmic division of the fifth cranial nerve.

The transparency of the cornea is attributable to its uniform cellular structure and its lack of vascularity (has no blood vessels) and detergescence. The transparency is maintained by the presence of a dehydrating pumping mechanism chiefly located in its endothelium and partly in its epithelium, which does not allow more than an optimum amount of water in its stroma. Once this dehydrating pumping mechanism is knocked out by sudden gross rise in intraocular pressure, as in acute congestive glaucoma or deep keratitis (tubercular or syphilitic) or an abnormal aqueous as in anterior uveitis, the water in the stroma reaches more than the optimum limit, making the cornea cloudy. Any ulceration deeper than the Bowman's membrane heals by laying down scar tissue; this causes corneal opacity that will interfere with vision if it is within the pupillary line. Superficial keratitis (corneal abrasion) anterior to the Bowman's membrane heals by migration of epithelial cells without any scarring, so no corneal opacity occurs. The cornea is normally bathed by the aqueous at its back and by tear film on its front.

The space between the back of the cornea and the front of the iris is called the anterior chamber and normally has aqueous that supplies nutrition to the cornea. The periphery of the anterior chamber is called the angle of the anterior chamber; its width depends upon the contour of the iris and the depth of the anterior chamber (see Chap. 2). The angle must have an optimum width to allow proper access of the aqueous to its microscopic outlet channels, which are present in the angle.

6.5 The Lens

The lens is a biconvex, avascular, colorless, and almost completely transparent structure, about 4 mm thick and 9 mm in diameter. It is suspended behind the iris by the zonule, which connects it with the ciliary body (Figs. 6.2 and 6.3). Anterior to it is the aqueous, and posterior to it is the vitreous (see Fig. 2.4). It has an anterior and a posterior surface, the posterior being more curved (Figs. 6.4 and 6.5). The central elevation on both surfaces is called the anterior and posterior poles (see Fig. 6.3). The lens capsule is a semipermeable membrane that controls an optimum amount of water inside. The capsule has a differential thickness and is thinner at the posterior pole (see Fig. 6.5). A subcapsular epithelium is present anteriorly (Fig. 6.6). The subcapsular lamellar fibers are continuously produced throughout life so that the lens gradually becomes larger and less elastic. The nucleus

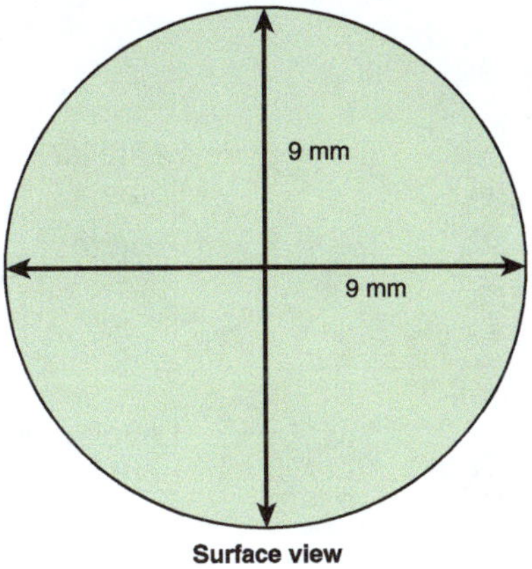

Fig. 6.2 Lens surface view

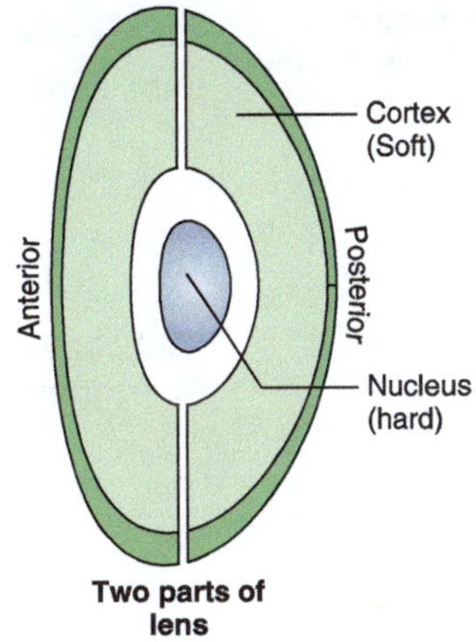

Fig. 6.4 Lens hard nucleus and soft cortex

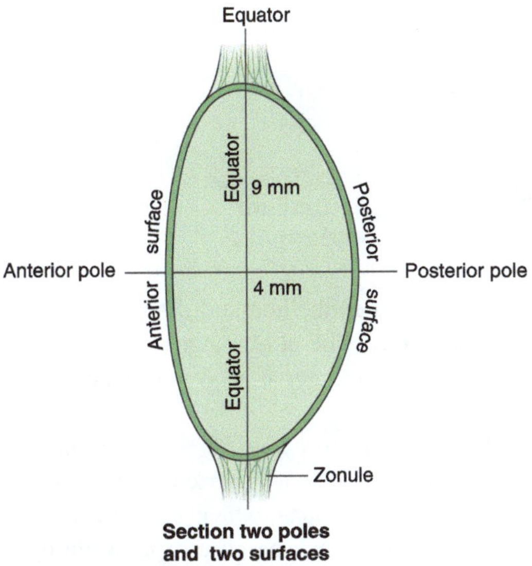

Fig. 6.3 Lens in section. Note the zonule at the equator and that the posterior pole is more curved than the anterior pole

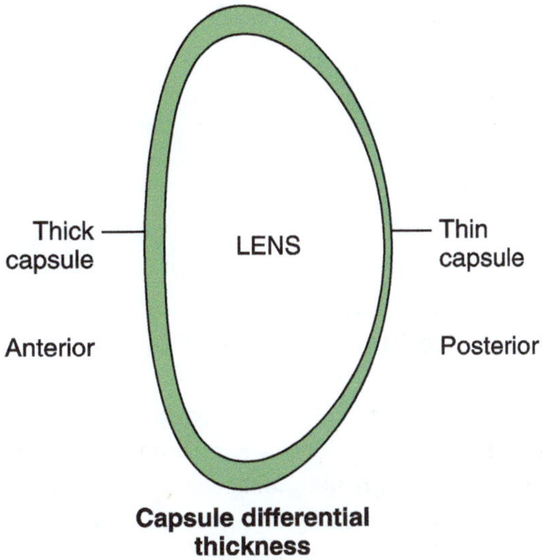

Fig. 6.5 Differential thickness of the capsule

and cortex (see Fig. 6.4) are made up of long, concentric lamellae; the nucleus is harder than the cortex. The suture lines formed by the end-to-end joining of these lamellar fibers are Y-shaped. When viewed with a slit lamp, the Y is upright anteriorly and inverted posteriorly. Each lamellar fiber contains a flattened nucleus (see Fig. 6.6). These nuclei are evident microscopically in the peripheral portion of the lens near the equator and are continuous with the subcapsular epithelium.

The lens is held in place by a suspensory ligament known as the zonule (zonule of Zinn), composed of numerous fibrils that arise from the surface of the ciliary body and insert into the lens matter. The lens is unique in that it is the richest organ in protein (35 %); it is 65 % water and a trace of minerals common to other body tissues. Potassium is more concentrated in the lens than in most tissues. Ascorbic acid and glutathione are present in both oxidized and reduced forms. There are no nerve fibers or blood vessels in the lens. Its nutrition comes from the aqueous, which bathes it on all sides.

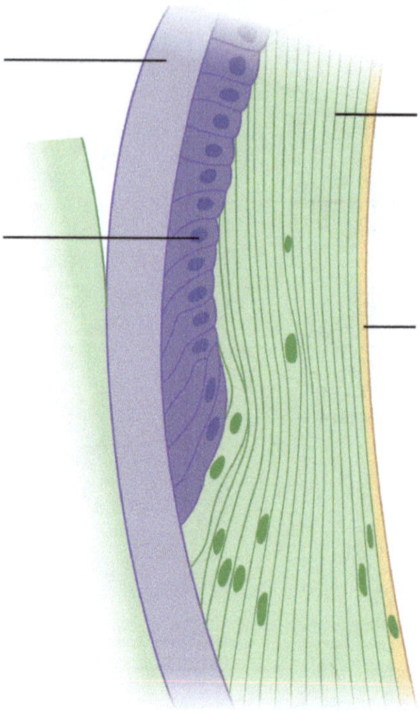

Fig. 6.6 Termination of the subcapsular epithelium; the lens grows throughout life

6.6 The Function of the Lens

Accommodation, a self-focusing device by which the human eye focuses on objects of interest from a reading distance of 14 inches to infinity, is achieved by an increased curvature of the anterior pole of the lens caused by an involuntary unstriated ciliary muscle present inside the ciliary body (see Chap. 2).

6.7 The Vitreous

The vitreous is a clear, avascular, gelatinous body that constitutes two-thirds of the volume and weight of eye, filling the space bounded by the lens, the retina, and the optic disc (Fig. 2.4, Chap. 2). The outer surface of the vitreous, the hyaloid membrane, is normally in contact with the following structures: the posterior lens capsule, the zonular fibers, the pars plana epithelium, the retina, and the optic disc. Its base maintains a firm attachment throughout life to the pars plana epithelium and the retina, behind the ora serrata. The attachment to the lens capsule and the optic nerve head is formed only in early life.

The vitreous is about 99 % water and the remaining 1 % includes two components—collagen and hyaluronan—which give it a gel-like form and consistency because of their ability to bind large volumes of water.

Suggested Reading

Parsons' Diseases of the Eye, edn 16. Revised by Miller SJH. London: Churchill Livingstone; 1979. pp. 1–14; 545–60

Riordon-Eva P, Cunningham E (2011) Vaughan and Asbury's general ophthalmology, edn 18. McGraw Hill, New York, pp 1–26

Trevor-Roper PD (1974) The eye and its disorders. Blackwell Scientific Publications, London, pp 26–44, 323–34

Wolff's E (1976) Anatomy of the eye and orbit, edn 7. Revised by Warwick R. London: HK Lewis; 1976; pp. 1–29; 30–180D

The Lacrimal Apparatus

7

Keywords

Lacrimal apparatus · Secretory parts · Drainage parts · Main lacrimal gland · Accessory lacrimal glands of Krause and Wolfring · Tear film · Lysozyme · Superior and inferior puncta · Vertical and horizontal canaliculi · Medial canthus · Lacrimal sac · Nasolacrimal duct · Inferior meatus of nose · Delayed canalization of nasolacrimal duct · Epiphora · Lacrimation

The lacrimal apparatus consists of two divisions. The first is the secretory part, which is responsible for the production of tears, and the second is the drainage part. The secretory part contains the main lacrimal gland, which is lodged in a fossa in the lateral part of the roof of the orbit, and two accessory lacrimal glands of Krause and Wolfring. The latter are present near the fornix. They are often called basal secretors because normally their secretions are sufficient to serve the purpose for which tears are meant without any need of drainage to the nose. The main lacrimal gland is brought into action when there is severe emotional stress or physical irritation caused by inflammation in the anterior segment of eye, causing an increase in the amount of secretion and a possible need for drainage to the nose. The afferent path for that reflex is via the ophthalmic division of the trigeminal nerve and the efferent path via the facial nerve. The main lacrimal gland is divided into a larger orbital and a smaller palpebral part by the lateral horn of the aponeurosis of the levator palpebrae superioris. The secretory ducts reach the lateral part of the superior fornix. One or two ducts may open directly into the inferior fornix.

The tear film has three layers. The outermost is the lipid layer from the meibomian glands which regulates the evaporation from the surface of the cornea and conjunctiva and prevents the spilling of tears on the face. The middle layer is aqueous, having soluble minerals and proteins that originate in the main and accessory lacrimal glands. The third innermost layer is mucoid and comes from the goblet cells of the conjunctiva. It contains glycoprotein mucin.

The tears have many functions. (1) They make the corneal surface smooth by covering the microscopic irregularities of the corneal epithelium, which is essential for adequate refraction at the cornea. (2) They keep the corneal and conjunctival surfaces moist for normal function. (3) They provide nutrition to the corneal and conjunctival epithelia. (4) They catch flying dust

M.W. Ansari and A. Nadeem, *Atlas of Ocular Anatomy*, DOI 10.1007/978-3-319-42781-2_7

particles and send them along the tear streak at the posterior sharp border of the lid margin. (5) They inhibit the growth of microorganisms by the antimicrobial action of lysozyme. (6) The constant evaporation of the tear film from the surface of cornea and conjunctiva takes heat from the conjunctival sac, allowing its temperature to be lower than that of other parts of body. This low temperature does not allow multiplication of microorganisms.

The drainage system is located in the upper and lower puncta at the junction of the ciliary and lacrimal parts of the lid margin, which are continuous with the vertical and horizontal parts of the canaliculus. The canaliculi open separately into the medial wall of the lacrimal sac housed in the lacrimal fossa. The lacrimal sac is continuous below with the nasolacrimal duct. The lower end of the nasolacrimal duct opens into the inferior meatus of the nose.

Secretary ducts from the main lacrimal gland open into the lateral part of the superior fornix and one or two open into the inferior fornix. The involuntary act of blinking spreads the tears like a swab in front of the cornea and conjunctiva; they also reach the posterior sharp border of the lid margins. This sharp posterior border of the upper and lower lid margins holds a horizontal streak of tears all along its length, which ends in the microscopic tear lake enclosed by the inner canthus into which the puncta dip. The capillarity and suction caused by the contraction of the orbicularis oculi in blinking make tears enter the puncta, canaliculi, and lacrimal sac. Once the tears reach the lacrimal sac, the elasticity of the lacrimal sac and gravity help them to reach the nasolacrimal duct from where they enter the inferior meatus of the nose. Drainage to the nose normally does not occur because of evaporation but only is seen during severe emotional stress or physical irritation caused by inflammation of the anterior segment of the eye when the main lacrimal gland comes into action.

7.1 Main and Accessory Lacrimal Glands

The structures responsible for the production and drainage of tears are collectively called the lacrimal apparatus. It has two divisions: the production of tears by the main lacrimal and accessory lacrimal glands and the drainage of tears.

The main gland is located in the lacrimal fossa, in the superior temporal quadrant of the orbit. This almond-shaped gland is divided by the lateral horn of the levator aponeurosis into a larger orbital and a smaller palpebral part. The smaller palpebral part can be visualized by everting the upper lid. The accessory lacrimal glands of Krause and Wolfring are located in the conjunctiva of the superior fornix and the superior tarsal border. They have the same structure as the main gland. They are known as basal secretors because their secretions are normally sufficient to maintain the health of the cornea. They lack ductules. The main gland comes into action in severe emotional situations, whereas routinely the secretions of the accessory glands are enough for normal functioning. There is no drainage to the nose in routine situations. Emotional or physical irritation caused by inflammation of the anterior segment may trigger an increase in secretion called lacrimation and may cause tears to flow copiously over the lid margin. The afferent pathway of the reflex arc is the ophthalmic branch of the trigeminal nerve. The efferent pathway is composed of parasympathetic and sympathetic contributions. Parasympathetic innervation starts from the pontine superior salivary nucleus and is conveyed by the greater petrosal nerve (contained in the separate part of the facial nerve and known as the nervus intermedius) and by the nerve of the pterygoid canal (vidian nerve), which synapses in the pterygopalatine ganglion and then moves via an uncertain route to the lacrimal gland. The sympathetic pathway is less well characterized. Because the gland is situated in the superior temporal portion of the orbit, its inflammation or dacryoadenitis will present as a painful red swelling in the lateral one third of the upper lid.

7.2 Components of the Tear Film

The surface of the eye is protected by a blinking reflex helped by the tears. The aqueous component dilutes the infectious material, mucus traps the debris, and the pumping action of the lids constantly flushes the tears to the tear duct. Tears contain antimicrobial substances, including lysozyme and antibodies.

The tear film (see Fig. 7.1) has three layers. The unicellular goblet cells of the conjunctiva secrete glycoprotein in the form of mucin, forming the innermost layer. The intermediate watery layer comes from the main and accessory lacrimal glands. The outermost lipid layer is produced by the meibomian glands of the tarsus.

1. The lipid layer is a monomolecular film thought to regulate evaporation and forms a watertight seal when the lids are closed.
2. The aqueous layer from main and accessory lacrimal glands contains water-soluble salts and proteins.
3. The deep mucinous layer overlying the corneal and conjunctival epithelial cells is composed of glycoprotein. Mucin is partly adsorbed onto the corneal epithelial cell membranes and is anchored by the microvilli of the surface epithelial cells. This provides a new hydrophilic surface for the aqueous tears to spread over, which becomes wetter with a lowering of surface tension.

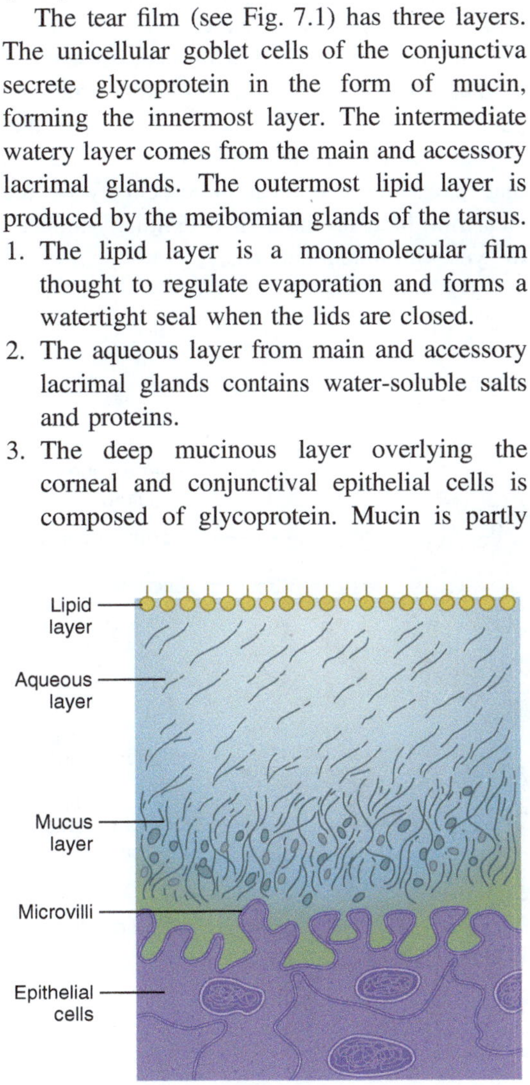

Fig. 7.1 Tear film covering the corneal epithelium

7.3 Composition of Tears and Volume

The normal tear volume is about 7 ± 2 μL in each eye. Albumin constitutes about 60 % of the total protein. Immunoglobulins IgA, IgG, and IgE as well as lysozymes make up the remaining 40 %. IgA predominates and differs from serum IgA in that it not only transudates from serum but is produced by plasma cells located in the lacrimal gland. In vernal conjunctivitis, the IgE concentration of tears increases. Tear lysozymes form 21 to 25 % of the total proteins and acting synergistically with gamma globulins and other non-lysozyme antibacterial factors, they represent an important defense mechanism against infection.

Potassium, sodium, and chloride also occur in higher concentrations in tears than in plasma. Tears have a small amount of glucose (5 mg/dL) and urea (0.04 mg/dL), and changes in blood concentration produces changes in tear glucose and urea levels. The average pH of tears is 7.35, although a wide normal variation exists (5.20–8.35). Under normal conditions, tear fluid is isotonic. Tear film osmolarity ranges from 295 to 309 mosm/L.

Dryness of the eye may result from any disease associated with deficiency of the tear film components (aqueous, mucin, or lipid), lid surface abnormalities, or epithelial abnormalities. Many of the causes of dry eye syndrome affect more than one component of the tear film or lead to ocular surface alterations that secondarily cause tear film instability. Dry spots may appear on the corneal and conjunctival epithelia. Filaments, loss of the goblet cells of the conjunctiva, increased cellular stratification, and increased keratinization may occur. The damaged corneal and conjunctival epithelial cells may stain with 1 % rose bengal.

7.4 The Schirmer Test

The Schirmer test is done with Whatman filter paper No. 41. The Schirmer strip is inserted into the lower conjunctival cul-de-sac at the junction of the middle and temporal thirds of the lower lid. The moistened exposed portion is measured without anesthesia after 5 min. Less than 15 mm wetting is considered abnormal. This test is a screening test of tear production. False-positive or false-negative results do occur.

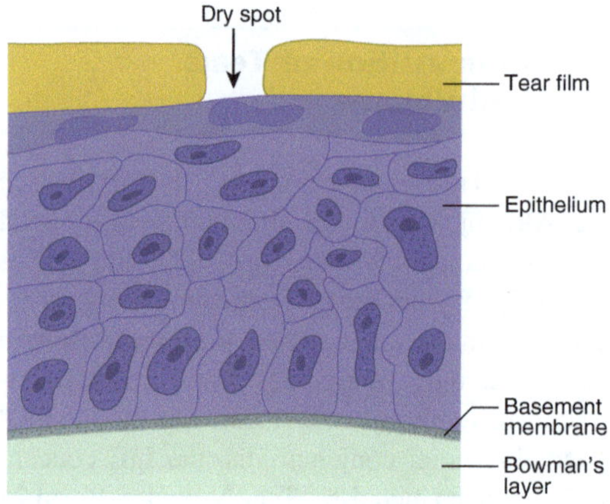

Fig. 7.2 Dry spot in tear film

7.5 Tear Film Breakup Time

The normal time is 15 s. Measurement of the tear film breakup time may be useful to estimate the mucin content of tears.

For the method, apply a slightly moistened fluorescein strip to the bulbar conjunctiva and ask the patient to blink. The tear film is then scanned with a cobalt filter on the slit lamp while the patient refrains from blinking. The time that elapses before the first dry spot (Figs. 7.2 and 7.3) that appears in the corneal fluorescein layer is the "tear film breakup time." It is definitely reduced in eyes with mucin deficiency.

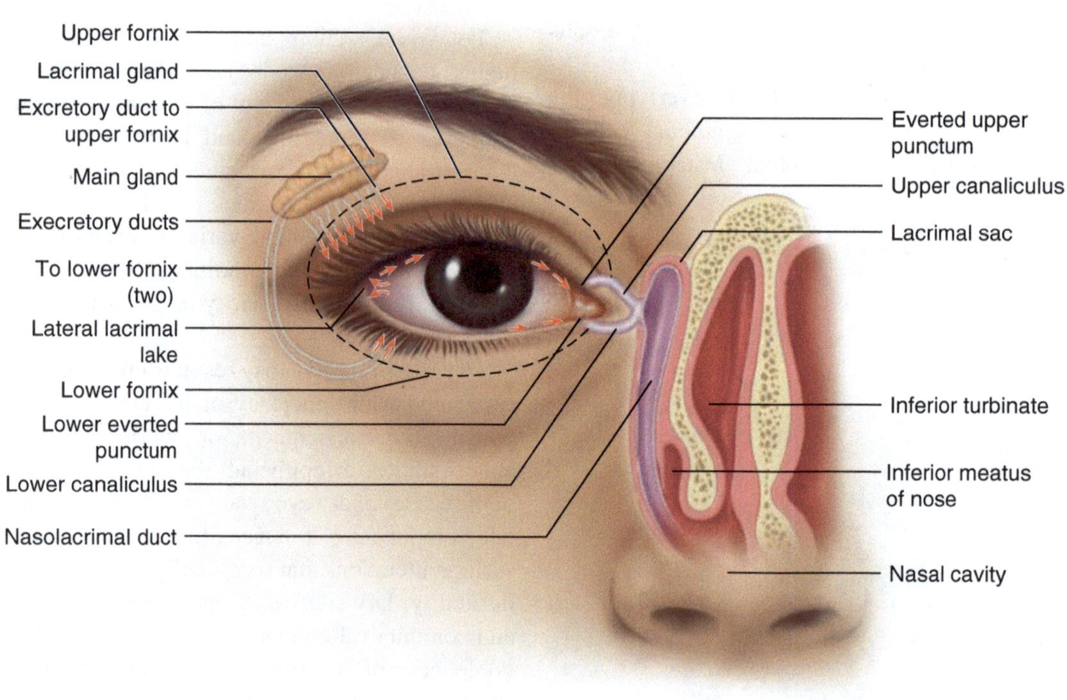

Fig. 7.3 Lacrimal apparatus: secretory part and drainage parts

7.6 Drainage Channels for Tears

The punctum is the hole present in a papilla at the junction of the ciliary and nonciliary (lacrimal) parts of the lid margins. The superior one is located in the upper lid and the inferior one is in the lower lid. Both puncta point inward and medially toward the globe dipping into the microscopic tear lake enclosed in the inner canthus (Figs. 7.3, 7.4 and 7.5). The lower canaliculus is 6.5 mm and is longer than the upper one, which is 6.0 mm. Both open separately through the lacrimal fascia into the medial wall of the lacrimal sac (see Fig. 7.5). The lacrimal sac is lodged in the lacrimal fossa in the medial wall of the orbit between the anterior and posterior lacrimal crests. The vertical and horizontal parts of the canaliculus are present in the lacrimal part of the lid margin, which encloses the inner canthus. The medial palpebral ligament is the Y-shaped ligament. The two limbs of Y enclose the lacrimal parts of the lid margin and have the canaliculi inside them. The anterior surface of the ligament is free and adherent to the skin and faces forward and laterally and the two branches look forward and medially, thus making an obtuse angle open in the forward direction. The medial palpebral ligament forms a well-marked prominence when the lateral canthus is pulled laterally and upward. It can be felt as a horizontal chord just medial to the inner canthus. The close relationship of the medial palpebral ligament with the lacrimal sac is shown in Fig. 7.5. The lacrimal sac makes an angle with the upper end of the nasolacrimal duct. That explains why that angle is the common site of blockage of the lacrimal passages. If you put a small finger just below the inner canthus, it goes into the lacrimal fossa. Digital pressure is used as a regurgitation test when mucoid discharge can be seen through the upper or lower punctum in cases of chronic dacryocystitis. Acute dacryocystitis presents clinically as a painful red swelling just below the inner canthus or lacrimal fistula also can occur in complicated cases.

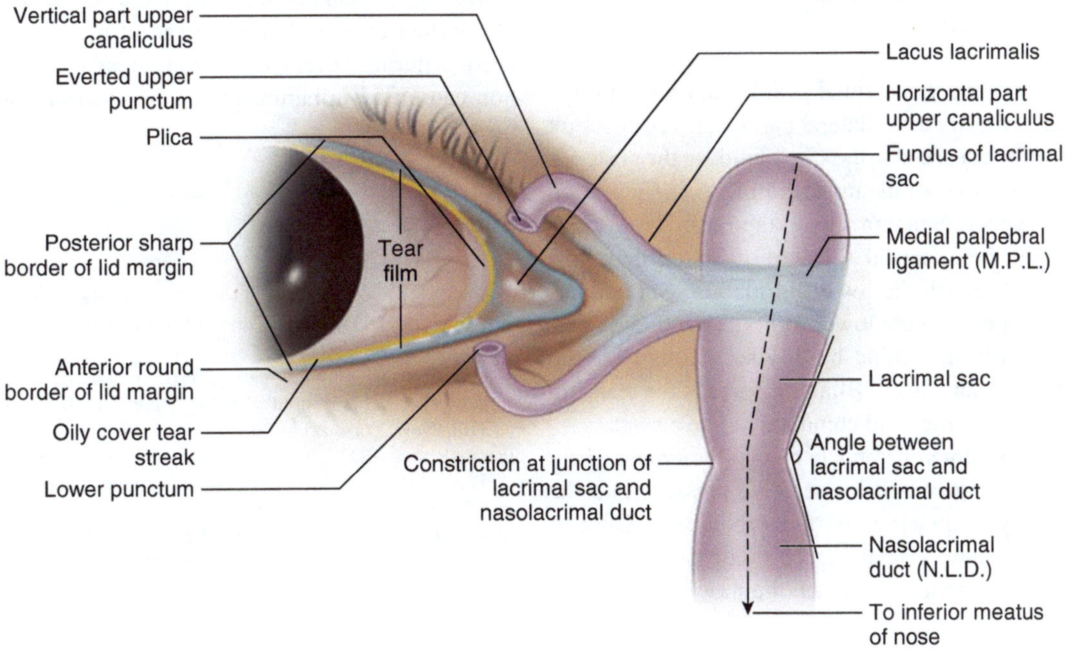

Fig. 7.4 Details of the inner canthus. Note the puncta, seen only after everting the lids. The *vertical* and *horizontal* parts of the canaliculus are seen opening separately into the lacrimal sac, which is continuous below with the nasolacrimal duct. Note the Y-shaped medial palpebral ligament

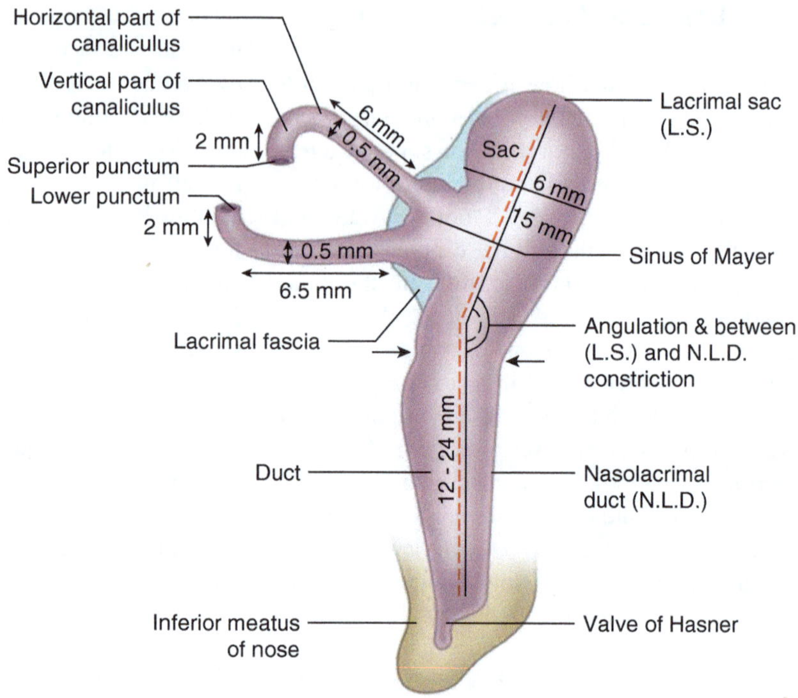

Fig. 7.5 Angulation and slight constriction at the junction of the lacrimal sac with the nasolacrimal duct, which explains its common site of obstruction

7.7 Conduction of the Lacrimal Fluid

Secreted by the lacrimal gland, tears pass through ducts mostly to the lateral part of the upper fornix and then descend to the strip of fluid at the posterior sharp border of the upper lid margin. They reach the upper punctum along this strip or directly under the upper lid. Closure of the lid may help them to reach the lower lid margin. Some ducts also open into the lower fornix. Tear fluid is prevented from spilling by oily secretions from the meibomian glands. Blinking spreads the tear film over the cornea and conjunctiva like a swab.

Tears enter the puncta by capillarity and partly by sucking caused by negative pressure into the sac by contraction of the orbicularis. Once it reaches the lacrimal sac, elasticity of the sac and gravity help them to reach the nasolacrimal duct, which opens into the inferior meatus of the nose. Normally, the tears do not reach the nose because of constant evaporation. Drainage of the nose is needed during severe emotional stress or irritation caused by inflammation of the anterior segment of the eye.

Suggested Reading

Parsons' Diseases of the Eye, edn 16. Revised by Miller SJH. London: Churchill Livingstone; 1979. pp. 1–14; 545–60

Riordon-Eva P, Cunningham E (2011) Vaughan and Asbury's general ophthalmology, edn 18. McGraw Hill, New York, pp 1–26

Trevor-Roper PD (1974) The eye and its disorders. Blackwell Scientific Publications, London, pp 26–44, 323–34

Wolff's E (1976) Anatomy of the eye and orbit, edn 7. Revised by Warwick R. London: HK Lewis; 1976; pp. 1–29; 30–180D

Neuro-ophthalmology: Neuromuscular Control of the Eyeball

8

Keywords

Optic nerve • Axons of ganglion cells of retina • Posterior scleral foramen • Optic chiasma • Optic tract • Lateral geniculate nucleus • Pupillary fibers • Pretectal nucleus • Neuromuscular control • Optic radiation • Visual center in occipital lobe • Nucleus of third, fourth, fifth, sixth, and seventh cranial nerves • Origin in posterior cranial fossa

The word *optic nerve* (ON) is a misnomer because this entity is not a nerve in the classic sense. Embryologically, it is a part of the brain, with which it has many common denominators. Both have a double blood supply, a peripheral and a central one, and it does not have a neurilemma; therefore after degeneration its fibers do not regenerate. It is likely to suffer from a degenerative disorder of the brain. Actually, it is a tract that consists of the axons of the ganglion cells of the retina that end in the lateral geniculate nucleus, where a new neuron takes over from its six layers. This new neuron's axons continue as the optic radiation, which ends in the cells of the visual center in the occipital lobe. The ON nerve has various parts: (1) a canalicular part inside the optic canal that measures 6 mm, (2) the cranial part (1 cm), (3) the orbital part (3 cm), and (4) the intraocular part (0.7 mm). Roughly, the total length of the ON is about 5 cm. The intraocular part anterior to the lamina cribrosa is about 1.5 mm wide, but as soon as it comes out, its fibers become medullated so that its diameter just behind the lamina cribrosa is about 3 mm. The orbital part of the nerve is continuous with the part inside the optic canal called the canalicular part. The canalicular part is accompanied by the ophthalmic artery, which crosses the undersurface of the ON from the medial to the lateral side. After this crossing the ophthalmic artery reaches the orbital opening of the optic canal, where it releases a branch called the central retinal artery. This artery travels on the undersurface of the sheaths of the ON before it pierces its three coverings to enter its center about 15 mm behind the eyeball. From there it travels forward inside the ON, giving branches to its pial network and some collateral branches that supply its posterior part in the orbit.

After giving the central retinal artery branches to the ON, the ophthalmic artery enters the cone of muscles from its lateral side. After supplying

M.W. Ansari and A. Nadeem, *Atlas of Ocular Anatomy*, DOI 10.1007/978-3-319-42781-2_8

many branches, the artery crosses the upper surface of the ON from the lateral to the medial end. Thus the ophthalmic artery crosses the ON twice: once on its undersurface from the medial to the lateral side inside the optic canal, and once inside the cone of muscles. It proceeds forward inside the cone of muscles to ultimately end by dividing into the dorsal nasal and supratrochlear branches. The ophthalmic artery is the first intracranial branch of the internal carotid artery that arises from its fifth bend as soon as it comes out from the roof of the cavernous sinus. If we keep in mind that the orbital cavity and the middle cranial fossa communicate with each other via the optic canal, it will simplify structures in that area. As soon as the ON enters the middle cranial fossa via a cranial opening of the optic canal after a journey of 1 cm, it joins the other ON to form a four-sided structure called the optic chiasma, which roughly lies above the pituitary fossa that houses the pituitary gland. Naturally, any growth of the pituitary gland can easily press on the overlying chiasma to produce a field defect called bitemporal hemianopia. Inside the optic chiasma, nasal fibers of the right eye cross to the left side and the temporal fibers travel to the same side inside the optic tract, which is a continuation of the chiasma. The optic tract crosses the cerebral peduncle to reach the lateral geniculate body. The axons starting from new cells of the lateral geniculate nucleus (LGN) proceed inside the brain as the optic radiation, which finally ends in the cells of the visual center in the occipital lobe. Thus the ON, optic chiasma, optic tract, lateral geniculate body, optic radiation, and the visual center in the occipital lobe all together constitute what is called the visual pathway. Any pathology in the various parts of the visual pathway produces a typical kind of field of vision that can help in diagnosis.

8.1 Optic Nerve

As stated previously, the word ON is a misnomer because it does not resemble other nerves in the classic sense. Embryologically, it is a part of the brain having many common denominators with it such as a dual blood supply (peripheral and central); it also has no neurilemmal cells so that its fibers cannot regenerate after disease, it is surrounded by three meninges of the brain, and it is prone to similar degenerative diseases. It can be divided into four parts (Fig. 8.1): (1) the intraocular part (0.7 mm), (2) the orbital part (3 cm), (3) the canalicular part inside the optic canal (6 mm), and (4) the intracranial part (1 cm). Its total length is about 5 cm.

The ON nerve starts as a flattened band from the anterolateral angle of a somewhat quadrilateral chiasma, runs forward in the optic canal. Here it is accompanied by the ophthalmic branch of the internal carotid artery, which arises as the first intracranial branch from the fifth bend of the internal carotid artery as soon as it emerges from the cavernous sinus. As it becomes more oval in shape and acquires its dural covering, the ON traverses the optic canal (see Fig. 8.1) with the ophthalmic artery in its sheath. The ophthalmic artery reaches the orbital end of the optic canal, where it enters the cone of muscles to cross the orbital part of the ON superiorly from the lateral to the medial end. At its point of entrance into the orbit, it contributes its first orbital branch, called the central retinal artery, which travels on the undersurface of the orbital part of the ON inside its sheath up to 15 mm behind the eyeball. There it enters the ON after piercing the dural sheath; it bends forward, along with the arachnoid and pia, to the center of that part of the ON. It travels forward along with its vein to the physiologic cup of the optic disc, where it bifurcates, making five bends before reaching its target. Being the first orbital branch of the ophthalmic, central retinal artery is more prone to occlusion in disease.

8.2 Relations of the Optic Nerve

The relationships of the ON in the intracranial part are as follows: first, it lies above the diaphragma sellae, which covers the pituitary body; then it lies on the anterior portion of the cavernous sinus. Above the nerve, the anterior cerebral branch of the internal carotid artery crosses it superiorly from the lateral to the medial

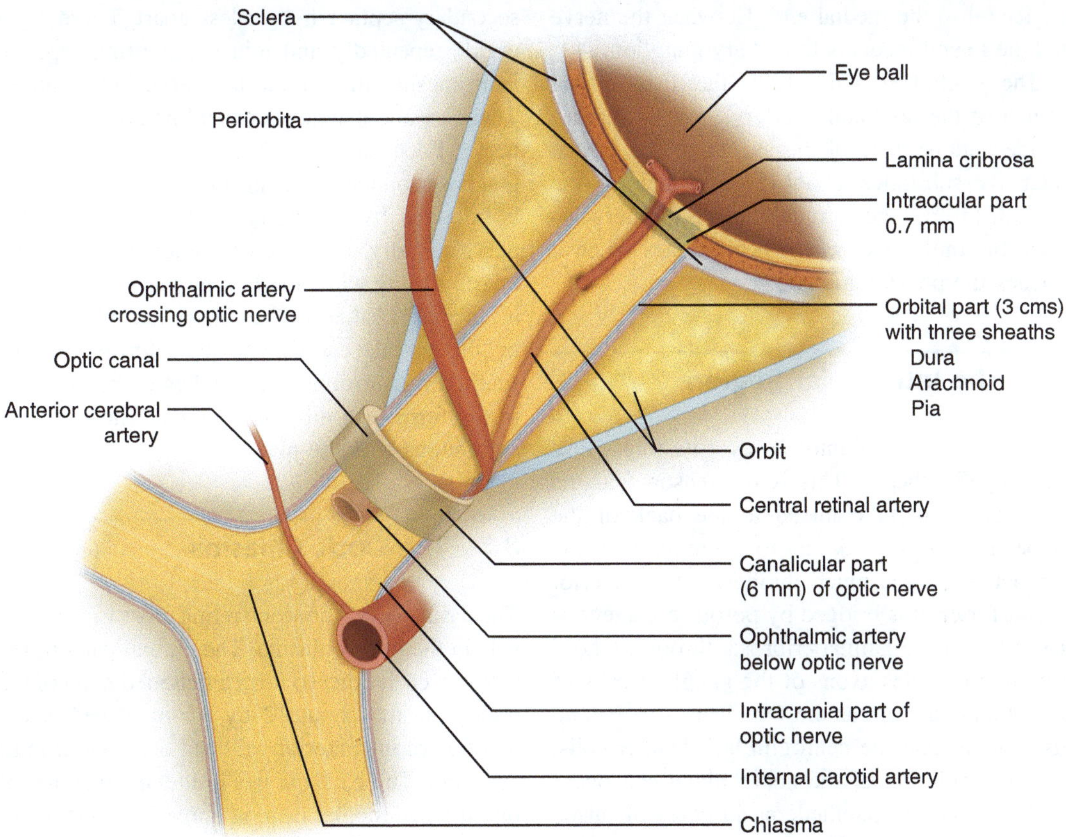

Fig. 8.1 Various parts of the optic nerve (*schematic*): (1) intraocular part (0.7 mm); (2) orbital part (3 cm); (3) canalicular part (6 mm); (4) intracranial part (1 cm). Note the three sheaths of the orbital part

end. The internal carotid artery is at first below and then lateral to it (see Fig. 8.1).

8.3 The Optic Canal

The pia forms a sheath that is closely adherent to the ON. The dura constitutes the periosteal lining to the canal, at the orbital end of which it splits to become continuous with the periorbita on the one hand and on the other with the dura of the ON (see Fig. 8.1). It should be noted that the subarachnoid space of the intracranial part of ON communicates with the subarachnoid space around the canalicular and orbital parts. The dura of the optic canal is attached to bone as well as to pia, thus giving the ON some degree of fixation in the canal. This points to the vulnerability of its blood supply in shear injuries of the ON. Medial to the ON is the sphenoidal sinus or a posterior ethmoidal sinus, from which it may be separated by a thin plate of bone. This explains the retrobulbar optic neuritis that may occur following a sinus infection.

8.4 The Orbit

At the optic foramen the nerve is surrounded by the origin of ocular muscles, those of the superior and medial recti being closely adherent to the dural sheath of the ON. This is the reason for the pain that occurs in extreme movements of the globe seen in retrobulbar optic neuritis. The nasociliary nerve, ophthalmic artery, and superior ophthalmic vein cross the nerve superiorly from the lateral to the medial end. Between the nerve and the lateral rectus is the ciliary ganglion.

The central retinal artery, the first orbital branch of the ophthalmic artery, travels forward on the undersurface of the sheath of the ON to enter its substance about 15 mm behind the eyeball. From there it travels inside the nerve to reach the optic disc, where it bifurcates when it reaches the physiologic cup.

8.5 The Intraocular Region

As the nerve passes into the eye, its fibers lose their myelin sheath. This is the reason that the nerve is 3 mm in diameter at the back of the globe and 1.5 mm adjacent to the retina (i.e., the intraocular part is much thinner). The posterior scleral foramen is bridged by porous collagenous tissue called the lamina cribrosa, through whose microscopic holes axons of the ganglion cells of the retina come out as the ON. This area has no visual cells, and the neuroglia lack Müller cells; therefore splitting of the nerve fibers can occur easily, leading to papilledema in cases of raised intracranial tension. This area as seen by direct ophthalmoscope is called the optic disc, in which bifurcation of the central retinal artery can be seen in the physiologic cup. The projection of the optic disc in the field of vision is called the "blind spot" because there are no rods and cones there; therefore this part of retina does not "see." The vessels from the vascular circle of Zinn (the short posterior ciliary arteries) supply the ON in that area. The sclera is continuous here with the outermost sheath of the ON nerve, the dura.

8.6 Structure of the Optic Nerve

The ON has about 1–13 million nerve fibers arranged in bundles divided by the septa. There are about six to nine very thick primary septa that divide the nerve into sectors. Between these sectors there are a great number of thinner secondary septa 1 mm or less apart. These septa divide repeatedly and join neighboring septa to form meshes that divide the nerve into bundles. There are about 40 septa extending into the whole nerve. Each one has a blood vessel at its center that is surrounded by collagen tissue, which again is bounded by neuroglia. The vessels enter the nerve radially, divide dichotomously repeatedly anastomosing with neighboring vessels, and form a vascular net that reaches the center of the nerve or the central vessels. The septa formed on this scaffolding surround the bundles of nerve fibers in the form of a tube or cylinder. The septa are continuous with the pia.

8.7 The Optic Chiasma

The nasal fibers from each retina cross to the other side in the optic chiasma. The cranial parts of the ON join each other to form a flattened band of the optic chiasma about 12 by 8 mm placed at the junction of the anterior wall and the floor of third ventricle. The chiasma lies in the anterior part of the interpeduncular cistern, above the diaphragm sellae. Inferiorly is the pituitary gland whose tumors can press the overlying crossing of nasal fibers of retina to cause bitemporal hemianopia. The nasal fibers of the retina cross to the opposite side in the chiasma, and the temporal fibers of the retina continue on the same side (Fig. 8.5). The hypophysis is located inferiorly and under its lateral edge is the cavernous sinus with its contents. The optic chiasma is continuous with the optic tract. Each optic tract has all the temporal fibers of the retina of the same side and all the nasal fibers of the opposite side. Eighty percent of the fibers in the optic tract are visual and 20 % are pupillary. The visual fibers continue along the whole of the optic tract to reach the nucleus of the lateral geniculate body. There the pupillary fibers separate themselves from the optic tract to reach the pretectal nucleus after decussation. From there a new nerve cell starts to reach the Edinger–Westphal nucleus of the third cranial nerve of the same and opposite side. That is how the third cranial nerve executes the parasympathetic supply to the sphincter pupillae through the ciliary ganglion.

This separation of the pupillary fibers from the optic tract indicates why in cerebral blindness, where the lesion is above the level of the lateral geniculate body, the pupils can be quite normal.

The pituitary gland is a small body about 12 × 8 mm in dimension. It is situated in the pituitary fossa called the sella turcica on the upper surface of the body of the sphenoid bone. The roof is formed by the dura mater and perforated at its center to allow the pituitary stalk or infundibulum to pass through. On each side it is flanked by the dura mater, which separates it from the cavernous sinus and its contents. In the lateral wall of the sinus from above downward are the oculomotor, trochlear, ophthalmic, and maxillary nerves (Fig. 8.2).

The optic tract (Fig. 8.3) crosses the cerebral peduncle, which has a pyramidal tract. A single lesion here can affect vision along the motor and sensory involvement. The fibers start from six layers of LGN to continue as the optic radiation. The radiation encompasses the posterior horn of the lateral ventricle forming the sinus of Meyer in order to reach the visual center in the occipital lobe. The optic radiation passes back in the white matter of the cerebral hemisphere and lies deep in the temporal lobe. It ends in the occipital lobe in an extensive area of thin cortex (1.4 mm or less in thickness) in which there is a distinctive white line.

8.8 Arrangement of Nerve Fibers in the Optic Nerve

Macular fibers are laterally placed in the distal part of the ON, which become central in the proximal part (Fig. 8.4). The figure shows the arrangement of the fibers in other parts, namely, the optic tract and the lateral geniculate nucleus.

In the human lateral geniculate nucleus, there are six well-defined layers of cells that are sharply separated almost throughout by the medullary lamina. These layers may be numbered one to six. The first two consist of large cells and four of the smaller cells. The crossed retinal fibers end in laminae one, four, and six, while the uncrossed fibers go to two, three, and five. Each optic fiber derived from a retinal ganglion cell ends within its appropriate lamina by dividing into five or six terminal branches, and each of these forms a small number of synaptic junctions with the dendrites of a single geniculate nerve cell. The smallest retinal lesion causes atrophy in all three layers belonging to that eye; hence the conducting unit in the ON is a three-fiber unit. The smallest lesion of the visual cortex causes atrophy in six layers of the lateral geniculate body, and so the conducting unit in the optic radiation is probably a six-fiber unit. The point-to-point projection of the retina to the lateral geniculate body is also carried from this body to the visual cortex.

8.9 The Optic Radiation

The fibers originating in the medial portion of the geniculate body and representing the upper portion of the retina form the upper portion of the optic radiation that moves to the upper lip of the calcarine fissure; those fibers coming from the lateral portion of the geniculate body form the lower portion of the radiation (Fig. 8.4). The macular fibers, like the ones in the geniculate body continue in the optic radiation to separate the fibers representing the upper portion of the retina from those representing the lower portion. This accounts for the fact that one may get a quadrantic visual field defect with a sharp horizontal border.

Fusion occurs only in the visual center of the visual cortex.

8.10 The Blood Supply of the Optic Nerve

All the arteries that supply the nerve tissue do so through the pial network of vessels. This network is rich and fine and extends to the back of the globe. In the intracranial portion of the nerve it is situated on the surface of the pia in the orbital portion in its thickness between the longitudinal

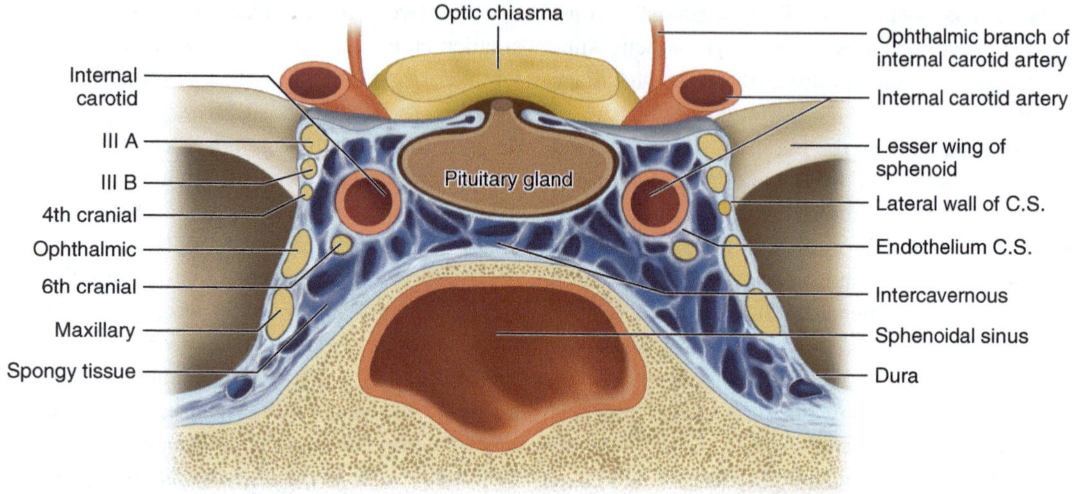

Fig. 8.2 The cavernous sinus. Note that the third (two branches) and fourth ophthalmic and maxillary nerves are in the lateral wall of the cavernous sinus. Note the origin of the ophthalmic artery as soon as the internal carotid artery pierces the roof from its fifth bend

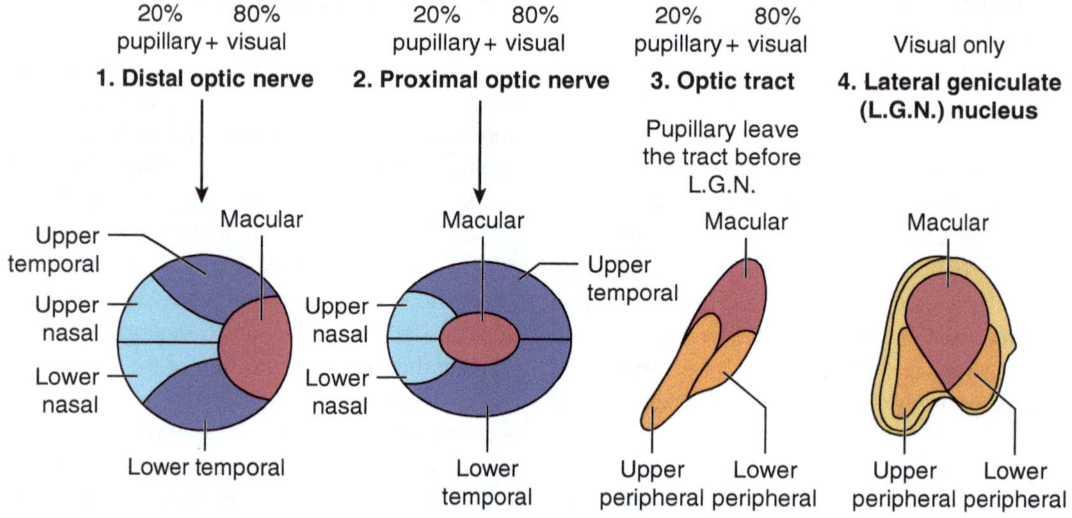

Fig. 8.3 Distribution of nerve fibers in the (1) distal and (2) proximal part of optic nerve, (3) optic tract, and (4) lateral geniculate nucleus

and circular fibers. There are two networks, one inside the other. The outer is the larger and is formed of arterioles of a fair size; the inner one lies within the first and consists of vessels so small that a loupe is necessary to see them. The network is supplied by arteries that probably anastomose slightly in the network but not before they reach it.

When the vessels enter the nerve they take with them a coat of pia and also a covering of glia, which constitutes the septa. The thickness of the septa is proportional to the size of the contained

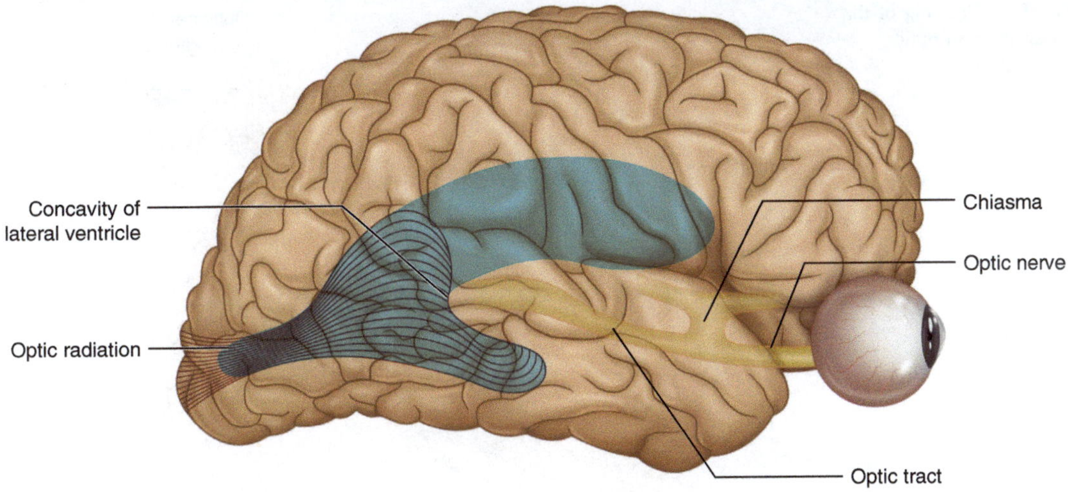

Fig. 8.4 Fibers of the optic radiation (nonpupillary)

vessel. The septa gradually disappear as they proceed toward the chiasma and the optic tract. The dura has very few blood vessels. It is the pial network, which acts as a distributing center, that provides a regular supply of blood to the nerve.

The vessels that join the pial network eventually come from the internal carotid artery. This is because the eye has grown out from the brain, its vessels have followed it. That explains why chiasmal vessels are short and those of the globe are larger.

8.11 The Intracranial Part

The feeders of the pial network are derived from (1) the ophthalmic artery, (2) the superior hypophyseal artery, (3) the anterior cerebral artery, and (4) the internal carotid artery. In the axial part of the intracranial portion, smaller vessels supply a larger number of fibers so that any interference with the blood supply will first affect the papillomacular fibers.

8.12 The Whole of the Remaining Part of the Optic Nerve

The remaining part of the ON is supplied by branches of the ophthalmic artery.

The intracanalicular part of the ON is supplied by the ophthalmic but differs from the orbital part in that the pial network is relatively poor. The branches involved are recurrent rami of the ophthalmic and central retinal arteries and the posterior ciliary arteries.

The orbital part of the ON is supplied by two groups: group A includes those that pierce the dura behind the entrance of the central retinal artery; group B includes those that enter the nerve or join the pial network at the site of entry.

Group A consists of six to twelve small vessels derived from the ophthalmic artery and the posterior half of its branches, including the central retinal artery. They reach the pial network after piercing the dura and the arachnoid.

Group B consists of the central collateral arteries, which pass backward to the optic foramen.

The region of the lamina cribrosa is supplied by the arterial circle of Zinn.

8.13 Chiasma (Fig. 8.5)

The chiasma is mainly supplied by anterior cerebral and internal carotid arteries helped by feeders from superior hypophyseal, anterior, and

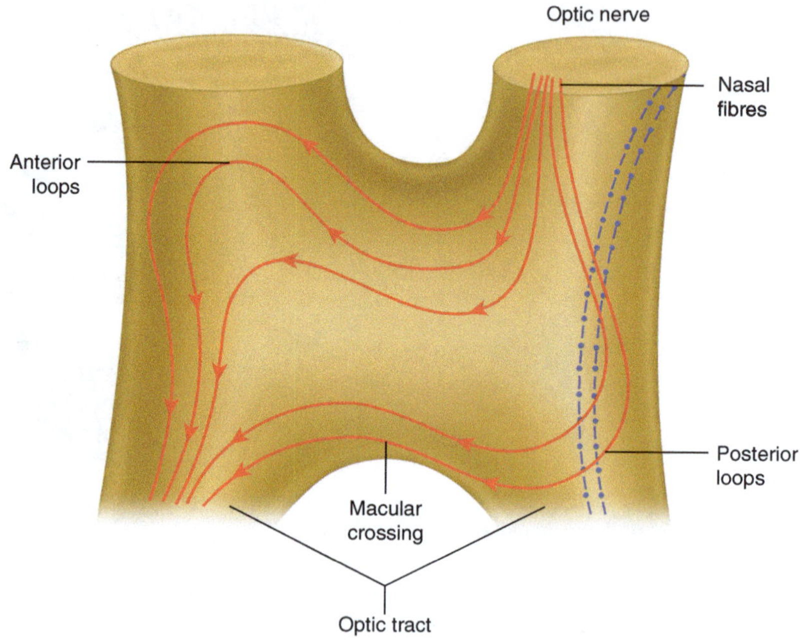

Fig. 8.5 Crossing of the nasal fibers in optic chiasma

posterior communicating and some from middle cerebral arteries. The anterior cerebral arteries supply most of the superior part of the chiasma, while the internal carotid arteries are mainly for the undersurface. The feeding vessels pass to the pial network and then to the chiasma.

8.14 Blood Supply of the Optic Tract

The feeder of the pial network comes from the posterior communicating artery partly but mainly from the anterior choroidal artery, which gives several branches to the tract. Some of the branches supply a portion of the optic radiation.

8.15 The Blood Supply of the Optic Radiation

The blood supply of the optic radiation originates from feeders of the pial network from perforating branches of the anterior choroidal vessels, from the deep optic branch of the middle cerebral and perforating cortical vessels, mainly from the calcarine branch of the posterior cerebral artery, and also from the middle cerebral artery.

The visual cortex receives feeders from the posterior cerebral artery via its calcarine branch.

8.16 Anatomic Correlation in Optic Atrophy

We are aware that the axon of one ganglion cell in the retina travels via the ON to the lateral geniculate nucleus, where a new synapse is established with cells of that nucleus. There are two ways that atrophy of ON fibers can occur. First, the parent ganglion cell of the ON fiber may be affected by retinal disease such as diffuse chorioretinitis or retinitis pigmentosa. Such a disease process will cause what is called the ascending type of optic atrophy in which atrophy starts at nerve head (the optic disc) and goes up via the ON up to the lateral geniculate nucleus.

Second, atrophy may occur when the ON fibers are pressed by a tumor or affected by other lesions during their course from the lateral geniculate nucleus to the optic disc. In such cases, atrophy will start at the site of pressure and take some time to reach the optic disc. This is technically known as the descending type of optic atrophy. For example, syphilitic arachnoiditis or direct pressure by a space-occupying tumor on the optic tract or chiasma causes atrophy of the ON fibers in that zone that takes some time to reach the optic disc. During this interval the optic disc may look normal for some time. Such optic atrophy is also called primary optic atrophy; no cause is visible in the retina but the optic disc becomes pale. Glaucomatous optic atrophy is caused by kinking of ON fibers at the lamina cribrosa and sclerosis of the capillaries supplying the optic disc.

8.17 Neuromuscular Apparatus: Nuclei

If the eye is to be rapidly and accurately fixed upon any object so that its image is formed upon the fovea (the most sensitive part of the retina) and if the two eyes are to rotate in unison in every rotation so that binocular vision is attained, it is essential that their motility and coordination be served by an unusually accurate and responsive neuromuscular apparatus.

The extraocular muscles are supplied by nerves arising from their nuclei in the midbrain. Their action is coordinated by intermediate centers situated in this region by which reflex activities are governed. Finally, these intermediate centers are linked with the vestibular apparatus, whereby they become associated with the equilibrium reflexes and also with the cerebral cortex so that voluntary movements and participation in the higher reflex involving perception become possible. The oculomotor or third cranial nerve supplies all the extrinsic muscles except the lateral rectus and the superior oblique. The superior oblique is supplied by the fourth cranial nerve (trochlear), and the lateral rectus by the sixth cranial nerve (abducens).

8.18 Some Helpful Hints About the Third, Fourth, Fifth, and Sixth Cranial Nerves

All these nerves have their origin in the posterior cranial fossa and have to pass from there to the middle cranial fossa in the region of the cavernous sinus. All assemble inside the cavernous sinus and all leave it to enter the orbit through the superior orbital fissure to supply the orbital structures. Some pass through the annulus of Zinn and some above it. Those that supply structures inside the muscle cone have to pass through the annulus of Zinn (i.e., two divisions of the third cranial nasociliary and sixth cranial nerves). Those supplying structures outside the annulus of Zinn will pass just above it (the lacrimal, frontal, and trochlear nerves).

8.19 The Third, Fourth, and Sixth Cranial Nerves in the Posterior Fossa

The third cranial nerve arises in the oculomotor sulcus on the medial side of the cerebral peduncle (Figs. 8.6, 8.7 and 8.8). It lies between the posterior cerebral and superior cerebellar arteries (Fig. 8.9). The nucleus of the third cranial nerve is around the aqueduct, from which fibers proceed toward the basis pedunculi after passing through the red nucleus and the substantia nigra to emerge in the oculomotor sulcus on the medial side of the cerebral peduncle (Figs. 8.7 and 8.8). From there the third cranial nerve proceeds above and medial to the trochlear nerve below and lateral to the posterior communicating artery and crosses the undersurface of optic tract from the medial to the lateral aspect.

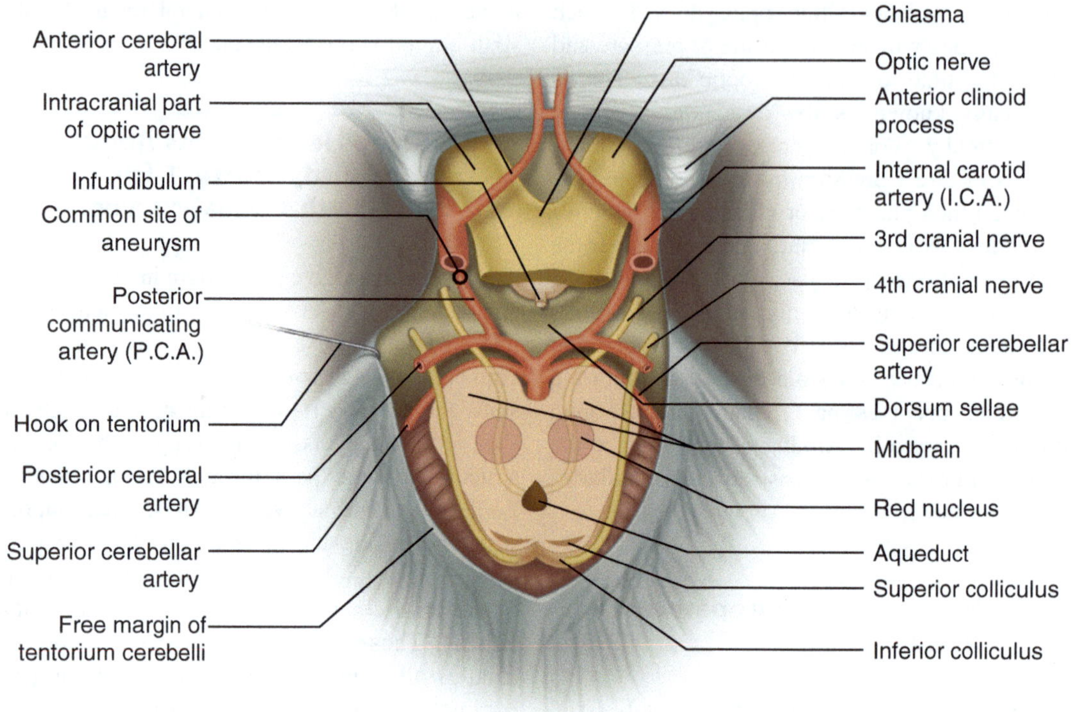

Fig. 8.6 Fibers of the third cranial nerve, starting at the aqueduct and then passing through the red nucleus. Also note the presence of third and fourth cranial nerves between the posterior cerebral and superior cerebellar arteries. Note the common site of aneurysm at the junction of posterior communicating and the internal carotid artery, which may cause third cranial nerve palsy because of its proximity

8.20 The Fourth Cranial Nerve (Trochlear)

The fourth cranial nerve (Figs. 8.10 and 8.11) is the most slender nerve with the longest intracranial course and the only nerve arising from the dorsal aspect of the central nervous system (Fig. 8.12). At the nuclear level fibers from one side cross to the other side to the superior medullary velum, which forms part of the roof of the fourth ventricle. Here it is crossed from below and upward by a branch of the superior cerebellar artery (Figs. 8.9 and 8.10). At the upper border of the pons, it runs forward between and parallel to the posterior cerebral and superior cerebellar arteries (Fig. 8.9). The trigeminal nerve, emerging from the lateral aspect of the pons just above its middle, passes forward below and lateral to the trochlear nerve. The oculomotor nerve here is above and medial to the trochlear nerve. Both proceed toward each other, and eventually the trochlear crosses the oculomotor nerve from lateral to medial (Figs. 8.13 and 8.14).

8.21 Sixth Cranial Nerve

The sixth cranial nerve in the posterior cranial fossa (PCF) emerges between the lower border of the pons and the lateral part of the pyramid (see Fig. 8.9); the basilar artery is in between the two. In Fig. 8.9 the dura is pierced at the back of the basilar portion of the occipital bone about 2 cm below and to the lateral side of the posterior clinoid process. The third, fourth, and trigeminal nerves are above it, gradually approaching the

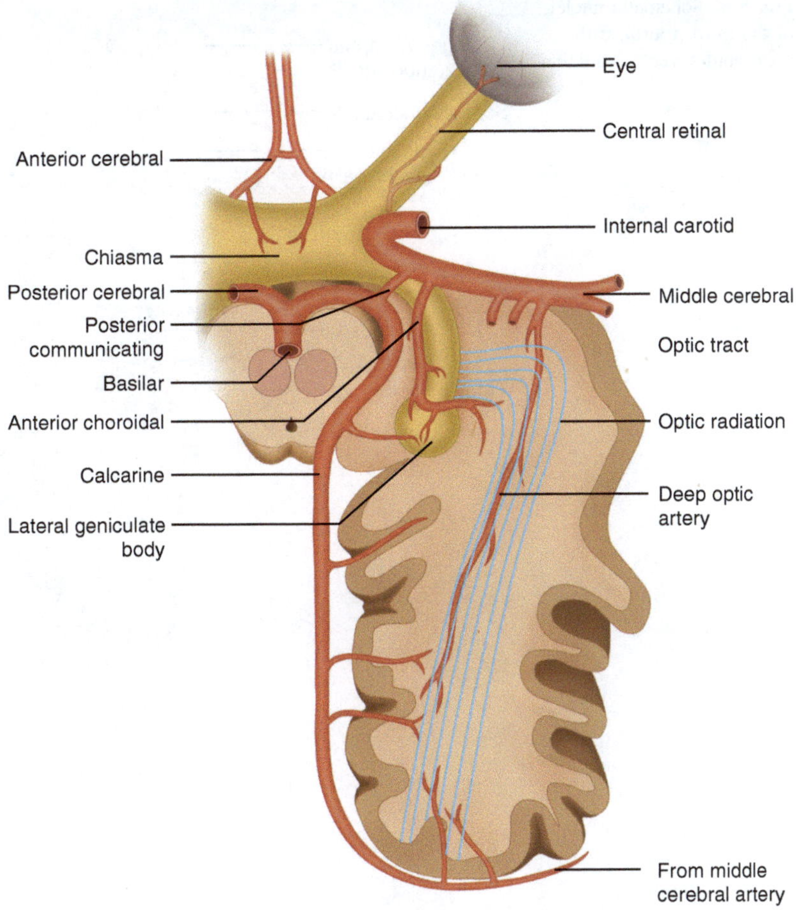

Fig. 8.7 Note the anterior choroidal, calcarine, and deep optic arteries supplying mainly the optic radiation

sixth. The sixth cranial nerve runs almost vertically at the back of the petrous temporal bone near its apex. At the sharp upper border of the petrous temporal bone the sixth cranial nerve bends forward practically at a right angle under the petrospheniodal ligamen. The third, fourth, and trigeminal nerves are above but are gradually approaching the sixth as they pass forward toward the middle cranial fossa, which is on its medial side on the petrosphenoidal ligament (Fig. 8.13). The third nerve is still medial to the fourth (see Figs. 8.13 and 8.14), and the sixth is trying to pierce the roof of the cavernous sinus (see Fig. 8.13). Thus we learn that the sixth cranial nerve enters the back part of the cavernous sinus along with the fourth, but the third cranial nerve enters the cavernous sinus through its roof. The ophthalmic division of the trigeminal nerve (Fig. 8.15) is on the lateral side heading for the superior orbital fissure. Just before the superior orbital fissure it divides into three parts: lacrimal, frontal, and nasociliary. The lacrimal, frontal, and trochlear nerves pass outside the annulus of Zinn because they have to supply structures outside the muscle cone. Two divisions of the third cranial, the nasociliary, and the sixth cranial nerves pass through the annulus of Zinn because they need to supply structures inside the muscle cone. Inside the muscle cone the ON is surrounded by nerves and branches of the ophthalmic artery, which crosses it superiorly. The ciliary ganglion is on the lateral side of

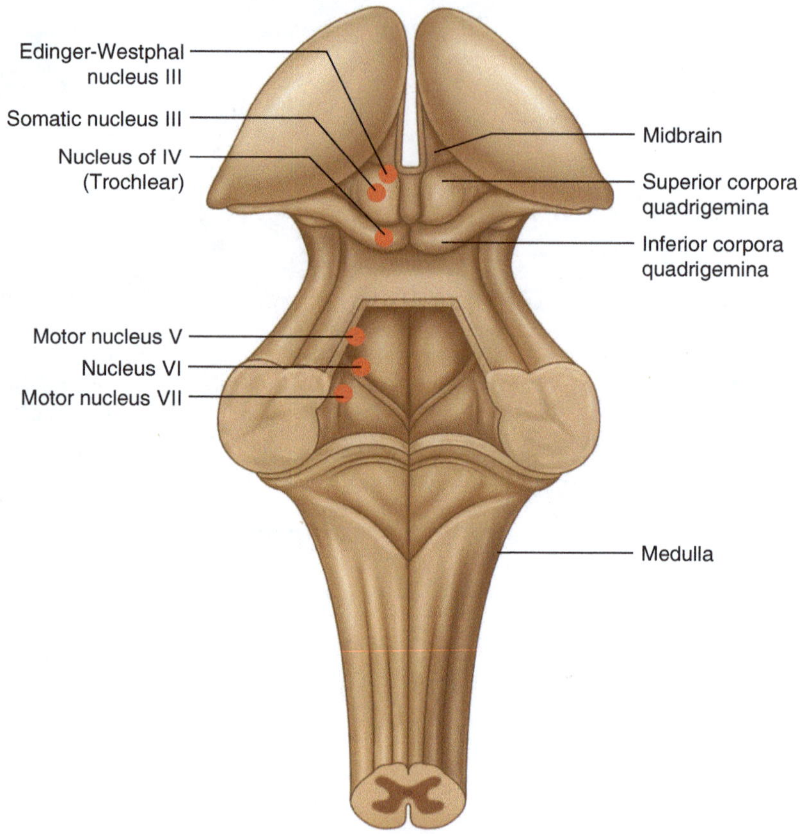

Fig. 8.8 Schematic nuclei of the third, fourth, fifth, sixth, and seventh cranial nerves

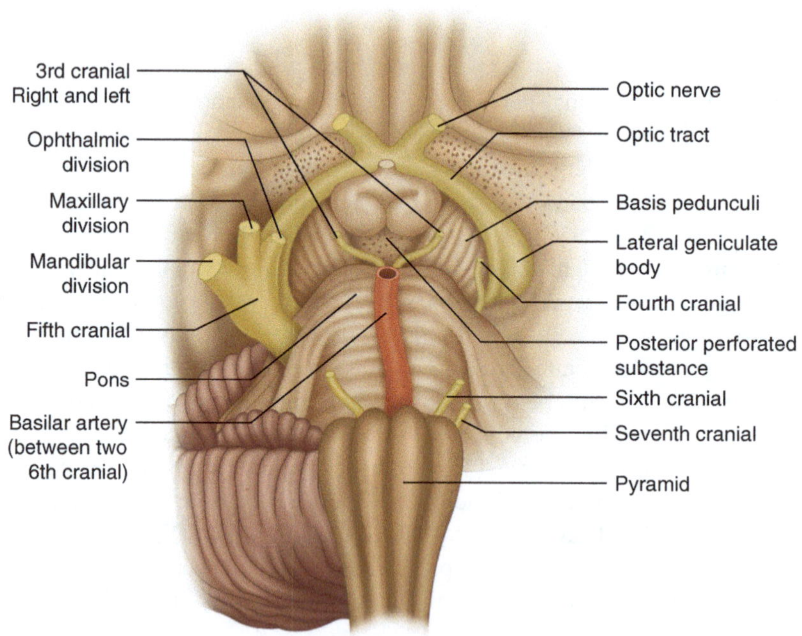

Fig. 8.9 Simplified origins of third, fourth, fifth, sixth, and seventh cranial nerves. Note the presence of the basilar artery between the right and left sixth cranial nerves

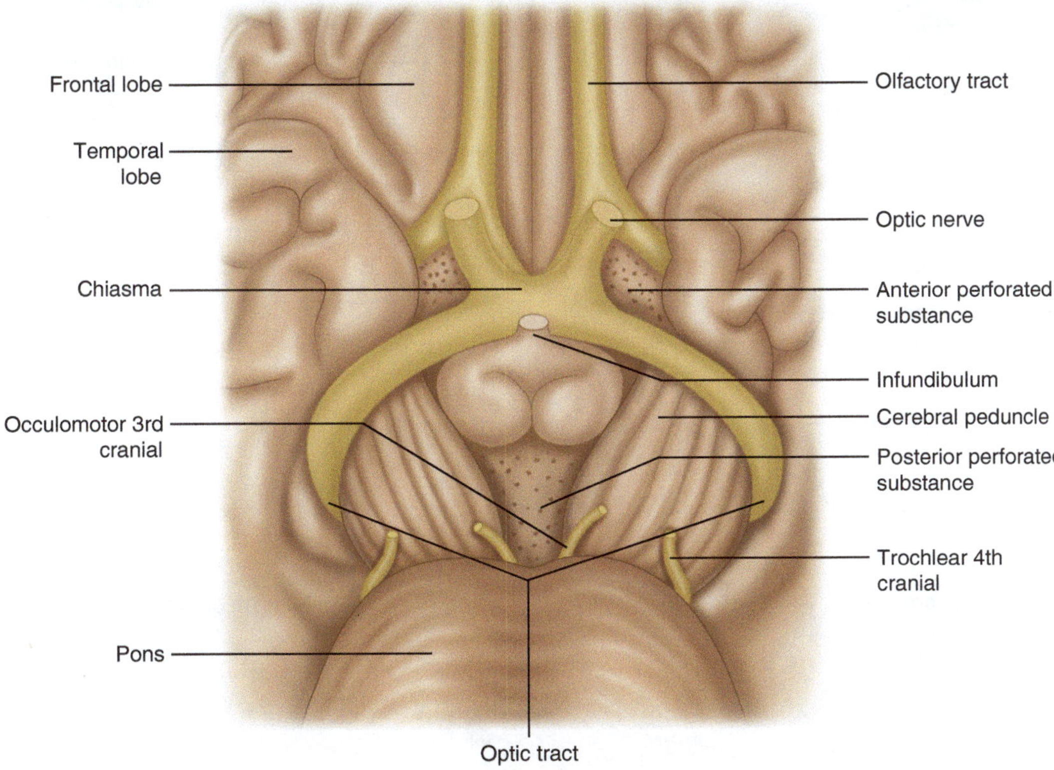

Fig. 8.10 Schematic origins of the third and fourth cranial nerves and relations of the optic chiasma

the ophthalmic artery and receives the motor root from the nerve to the inferior oblique. The short and long posterior ciliary arteries and nerves surround the ON inside the cone. The motor fibers from the third cranial nerve relay in the ciliary ganglion. A new neuron takes over to send short and long posterior ciliary nerves, which ultimately proceed to the sphincter pupillae and the ciliary muscle in the ciliary body. (the parasympathetic supply to pupil) (Fig. 8.16).

8.22 Arrangement in the Cavernous Sinus

The two divisions of the third and fourth cranial nerves and the ophthalmic and maxillary nerves are in the lateral wall of the cavernous sinus. The sixth cranial nerve lies below and lateral to the internal carotid artery enclosed separately in a membrane (see Fig. 8.2). The medial longitudinal bundle connects all the above nuclei of each side as well as regulates the conjugate movements of the eyes (Fig. 8.17).

8.23 Overall Review of the Visual Pathway

The visual pathway extends from the ON, chiasma tract, lateral geniculate body, and optic radiation to the visual center in the occipital lobe. Pathologies at its various parts show typical field defects (Fig. 8.18), especially with colored objects. Generally speaking, vascular lesions tend to produce steep borders, that is, the field defect is of the same size no matter what the intensity, size, or color of the object. The more

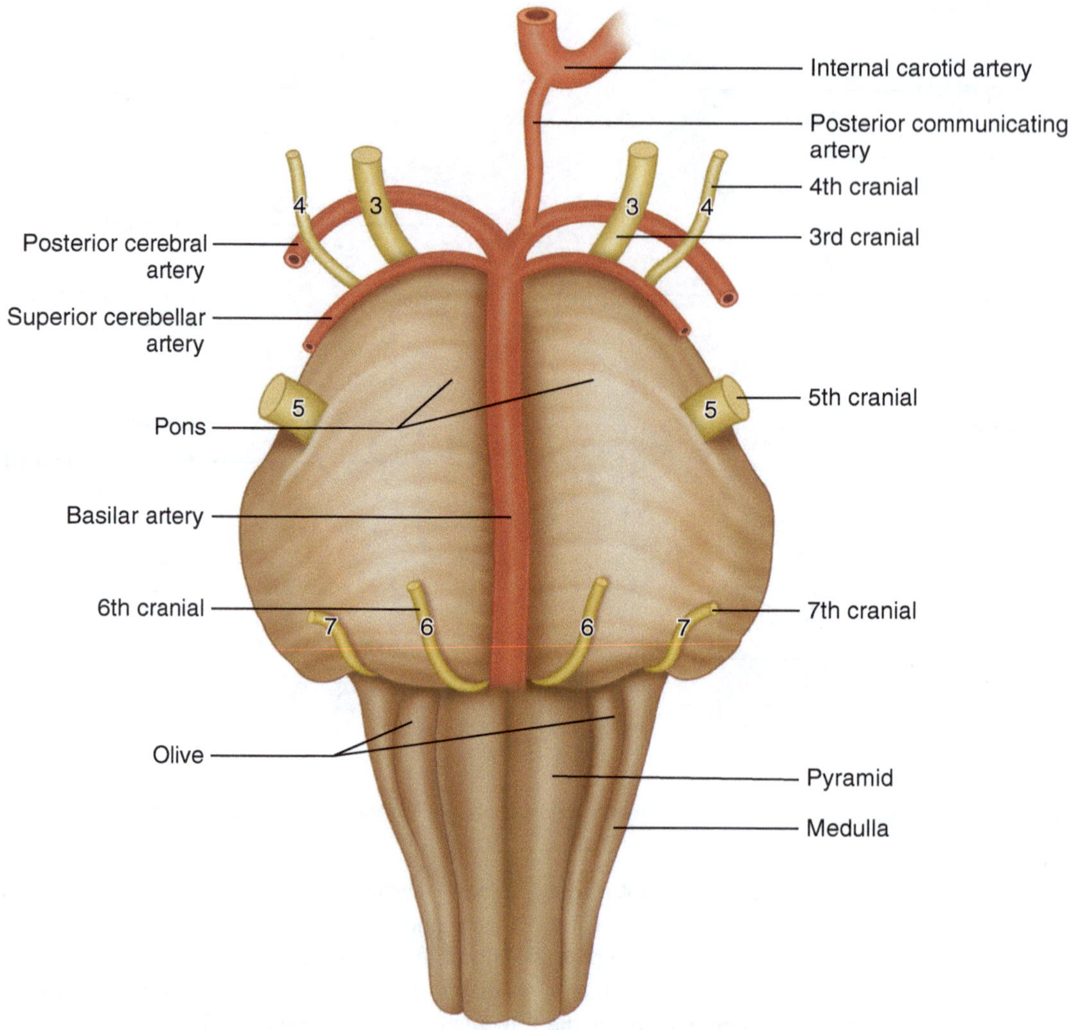

Fig. 8.11 Simplified schematic ventral view of the brainstem with origins of the third, fourth, fifth, sixth, and seventh cranial nerves. Note the presence of third and fourth between two arteries, the posterior cerebral and superior cerebellar

similar the defect is in size, shape, and location in the two eyes, the farther posterior the lesion is likely to be in the visual pathway. A lesion in the occipital region tends to cause identical defects in each field, whereas optic tract lesions tend to cause dissimilar homonymous field defects. Owing to the dual vascular supply to the occipital lobe from the middle and posterior cerebral circulation, occipital infarcts may spare the occipital pole; this is called macular sparing or macular splitting. Occipital lesions may also produce what is called residual sight, in which response to movement may be demonstrated in the hemianopia field in the absence of form sense, meaning poor visual acuity. When a complete homonymous hemianopia occurs, irrespective of the site of the lesion, there should still be intact visual acuity in each eye, since macular function is still present in the retained field. Central retinal vein occlusion, ocular hypotony, and intraocular

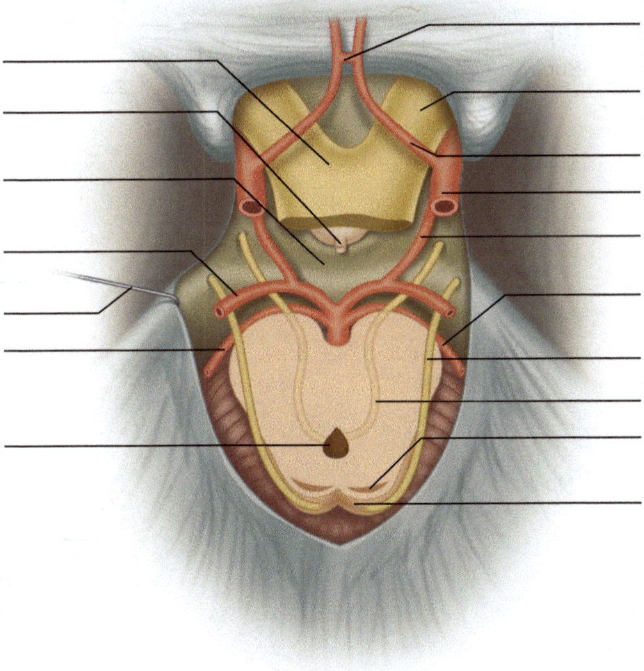

Fig. 8.12 Circle of Willis and optic chiasma

inflammation can produce optic disc swelling (papilloedema) and hence give the misleading impression that there is ON disease.

The criteria for ON disease are afferent pupillary defect, poor color vision, and optic disc changes. The optic disc may be normal in the early stages of descending optic atrophy. The lesion proceeds from the distal part of the ON after compression by an intracranial lesion. For the diagnosis of anterior ischemic optic neuropathy, optic disc swelling should be present in the acute stage. In general there is a correlation between the degree of optic disc pallor and the loss of acuity, visual field, color vision, and pupillary responses.

8.24 Optic Neuritis

Painful loss of vision (gross), poor color and contrast sensitivity, afferent pupillary defect, and diplopia are cardinal symptoms and are useful in differentiating papillitis from papilloedema. Painful sudden loss of vision in women is common, mostly in the fourth decade. Pain is caused by connection of the sheath of the ON near the optic foramen to the origin of the medial and superior recti. Diplopia is mostly caused by internuclear ophthalmoplegia.

Other causes of optic neuritis are viral in children and herpes zoster ophthalmicus.

The cause of anterior ischemic optic neuropathy is infarction of the retrolaminar part of the ON resulting in decreased perfusion of the short posterior ciliary arteries. It may also be caused by neuropathy attributable to giant cell arteritis, the symptoms of which are a painful tender temporal artery and myalgia. Confirmation is obtained by biopsy of the temporal artery.

8.25 Optic Nerve Compression

The possibility of ON nerve compression should be considered in any patient with visual loss not explained by an intraocular lesion. Intraorbital

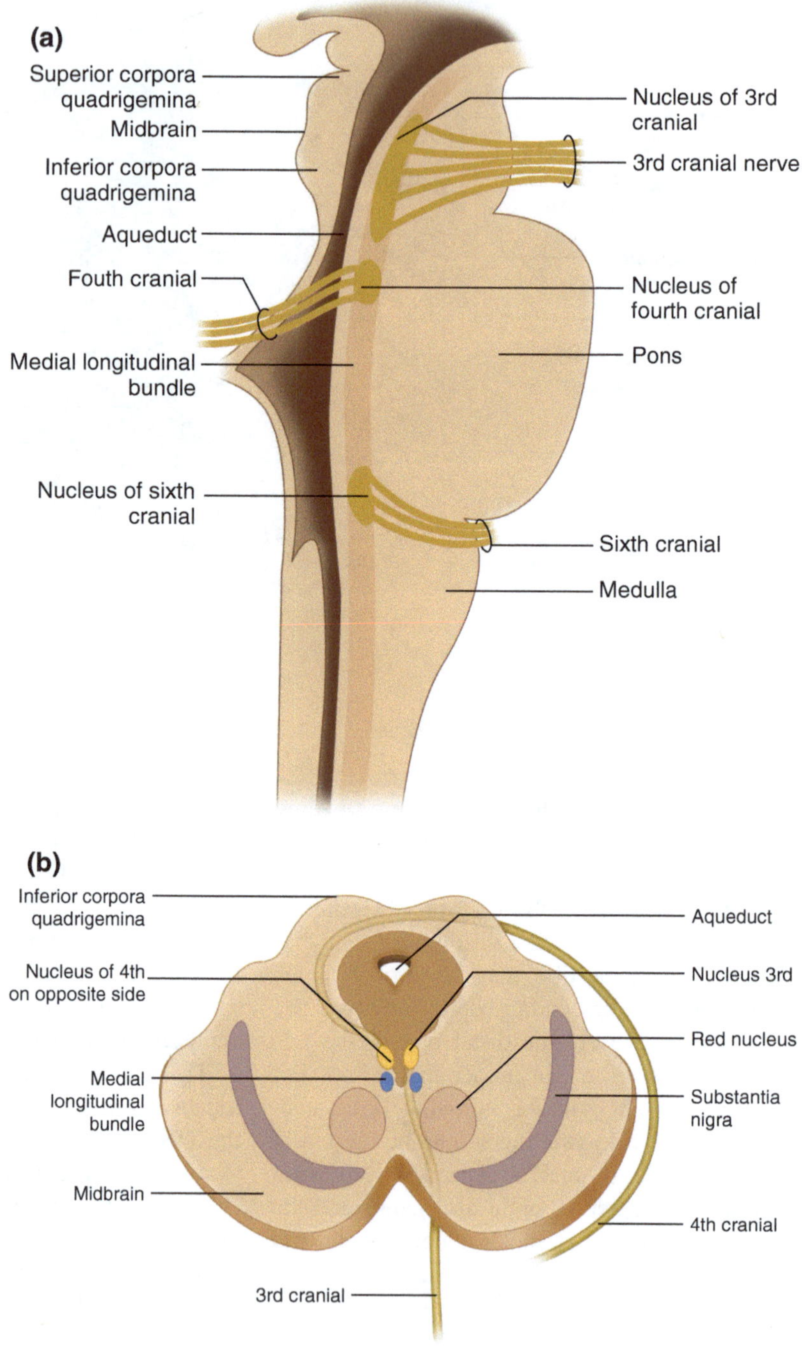

Fig. 8.13 **a** The origins of the third, fourth, and sixth cranial nerves from their nuclei. **b** Fibers of the fourth cranial nerve cross to the opposite side before its emergence while the third cranial nerve emerges to the same side

compression of the ON may show papilloedema early, but when the compression is intracranial, the optic disc may not show any changes until the descending optic atrophy reaches the disc. In any suspected cases of compression of the ON, early imaging by magnetic resonance imaging or computed tomographic scan will help.

Intracranial meningiomas that may compress the ON include those arising from the sphenoid wing, the suprasellar meningioma, and the

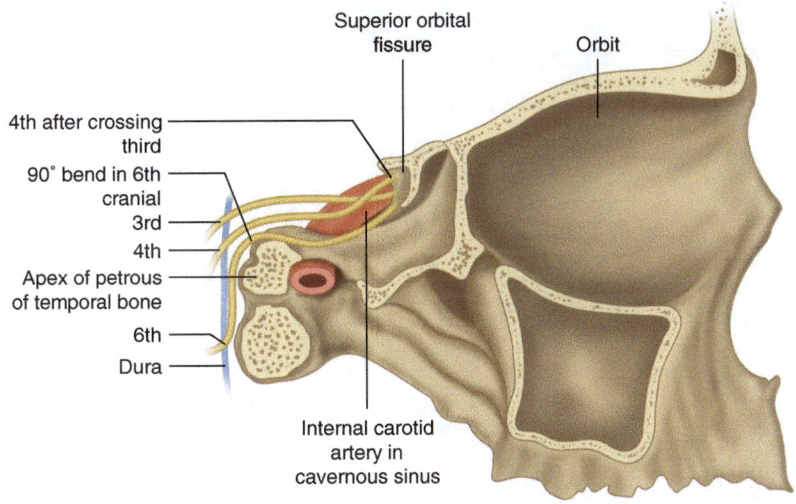

Fig. 8.14 A highly simplified illustration of 90° bend in the course of sixth cranial nerve at the apex of petrous part of temporal bone. Note also the crossing of the third by the fourth cranial nerve

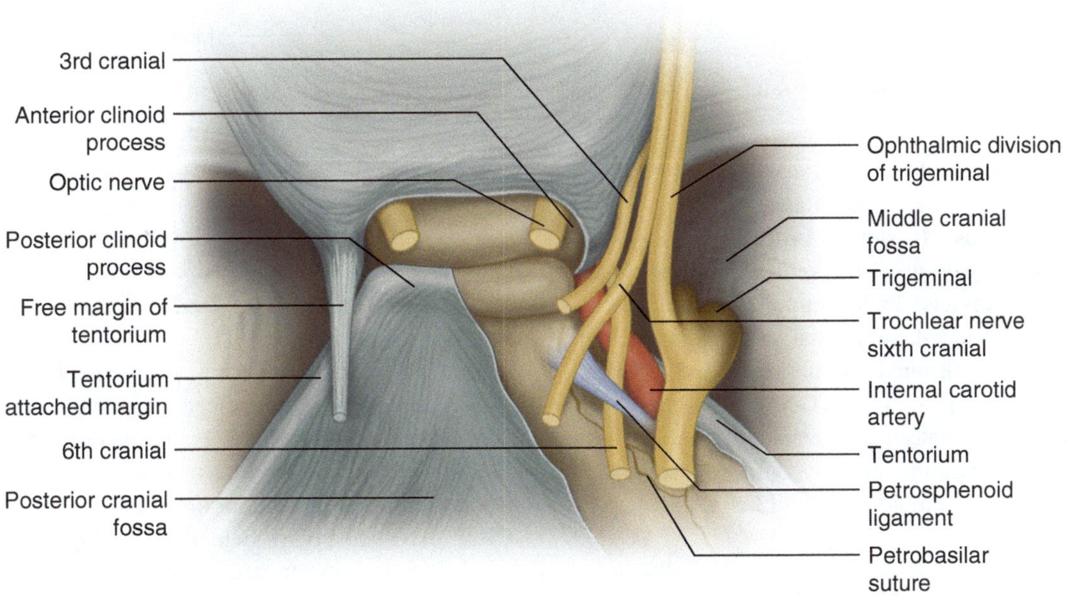

Fig. 8.15 The third, fourth, and sixth cranial nerves entering the cavernous sinus. Note that the fourth lies on the petrosphenoidal ligament, whereas the sixth lies deep in it

olfactory groove. In addition to field of vision changes, sphenoidal meningioma also produces proptosis, oblique proptosis, ocular motility disturbance, and trigeminal sensory loss. Primary meningiomas of the sheath of the ON are rare and occur mostly in middle-aged females.

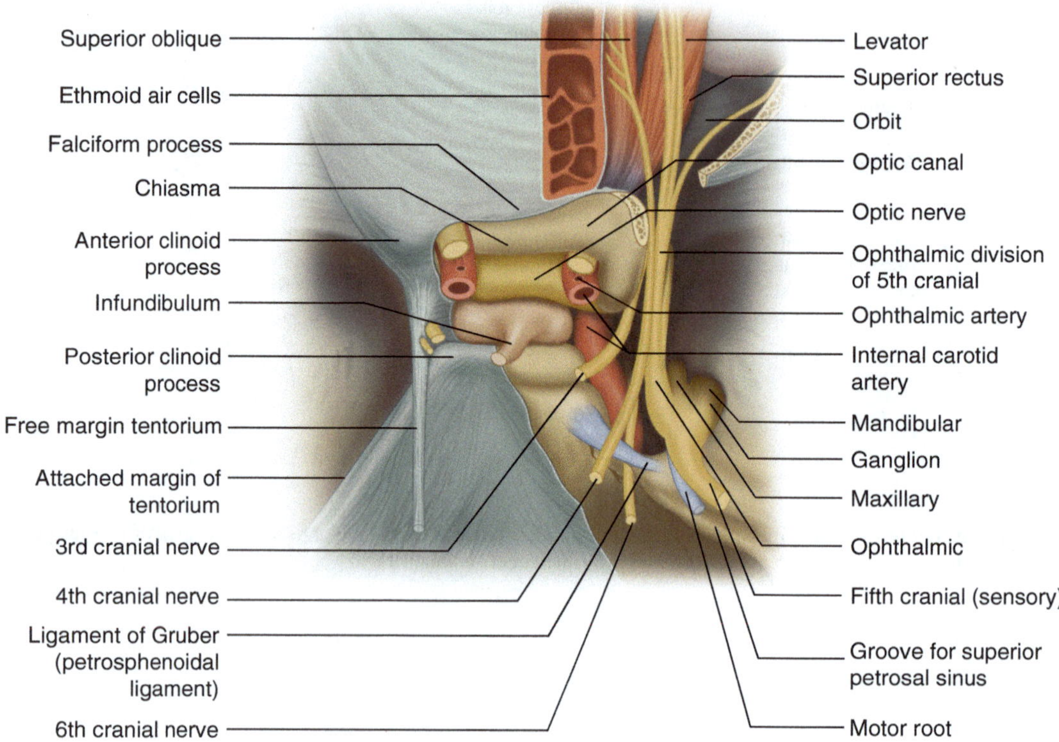

Fig. 8.16 Nerves entering the orbit from middle cranial fossa after leaving the cavernous sinus

Proptosis in these cases may be straightforward, and the changes are bilateral in 5 % of cases. Visual loss is slowly progressive, showing a swollen pale disc with retinochoroidal collateral blood vessels. In leukemia, non-Hodgkin lymphoma, and disseminated carcinoma, ON infiltration with gross visual loss and optic disc swelling may occur. ON involvement is an important cause of permanent visual loss in orbital cellulitis, which could be secondary to untreated ethmoiditis in children (see Chap. 3).

8.26 Optic Nerve Trauma

Direct ON injury occurs in penetrating orbital trauma, including local anesthetic injections for ocular surgery (retrobulbar injection) and in mid-facial fractures involving the optic canal. ON avulsion may result from an abrupt rotational injury to the globe, such as from being poked in the eye with a finger.

8.27 Lesions of the Optic Chiasma

Lesions of the optic chiasma may cause bitemporal hemianopia field defects (see Fig. 8.18). Central visual acuity may be preserved. Most diseases that affect the chiasma are neoplastic, such as tumor of the anterior lobe of the pituitary. There may be special field defects (Fig. 8.19) such as loss of vision and cranial nerve palsies because of the nearness of the chiasma to the cavernous sinus, which contains the third, fourth, and sixth cranial nerves.

Gliomas of the ON are rare and occur mostly in children under 7 years of age; they are accompanied by visual loss, proptosis strabismus, or nystagmus.

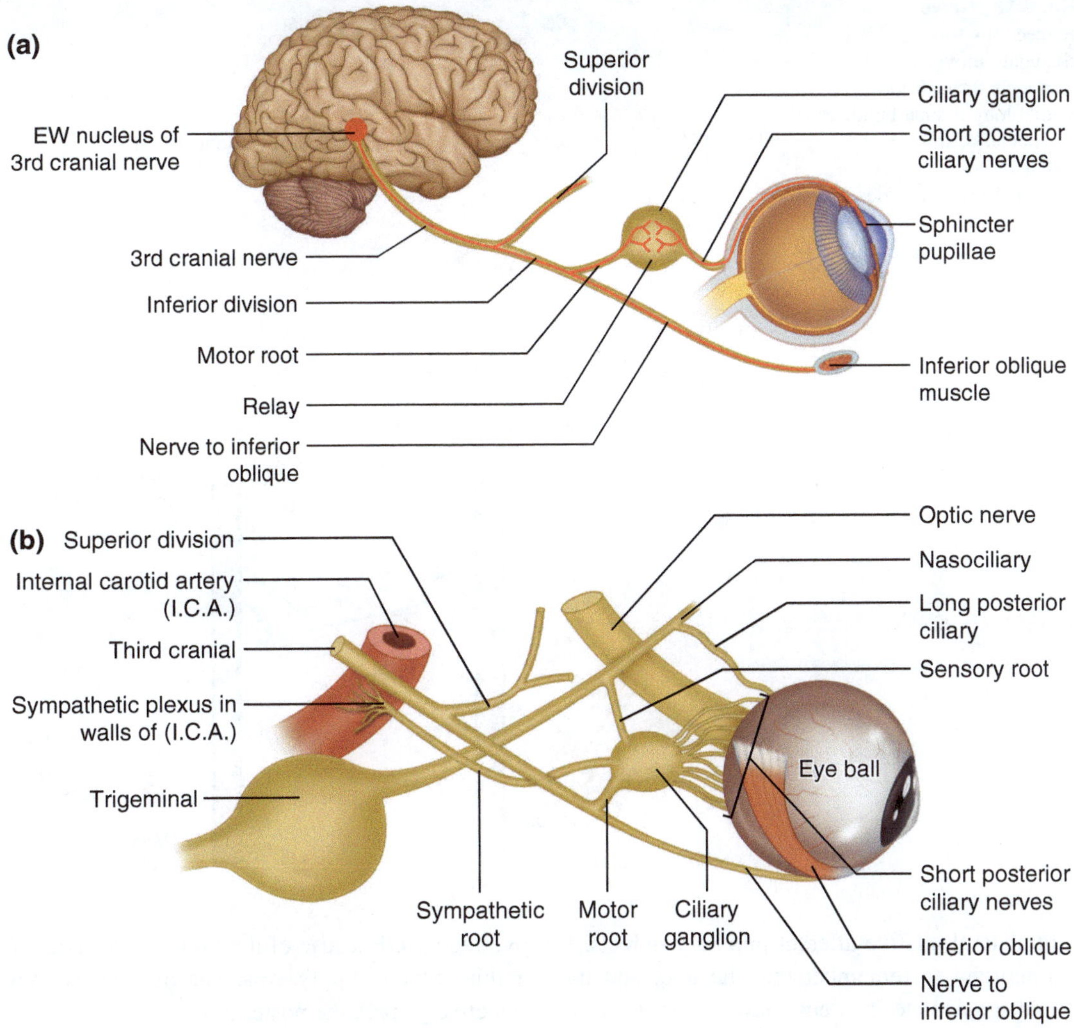

Fig. 8.17 **a** The parasympathetic nerve supply to the sphincter pupillae via the third cranial via ciliary ganglion. **b** Schematic roots and connections of ciliary ganglion

8.28 The Retrochiasmatic Visual Pathway

Cerebrovascular disease and tumors are responsible for most of the lesions. They produce contralateral homonymous field defects. Lesions of the optic tract, lateral geniculate nucleus, or optic radiation tend to produce dissimilar visual field defects with more involvement in the eye with the nasal defect. Posterior partial lesions in the optic radiation or occipital cortex tend to produce more similar visual field defects. Any unilateral tract or optic radiation lesion will spare visual acuity because the visual pathway in the other half of the brain is intact. Optic tract lesions or those of the lateral geniculate nucleus are uncommon. In tract lesions there may be

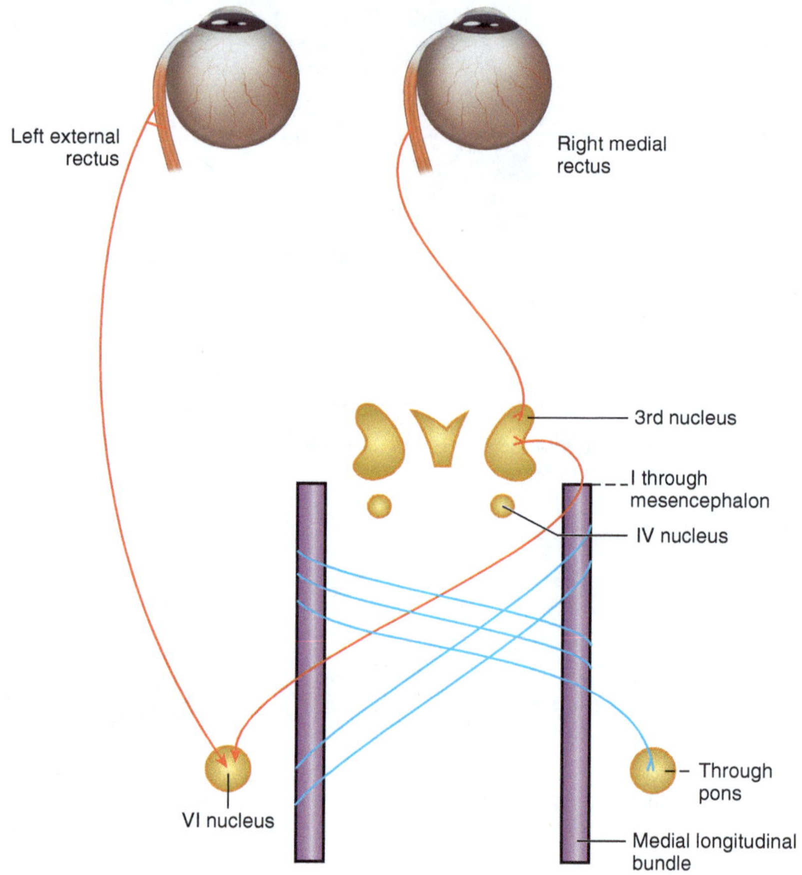

Fig. 8.18 Nerve connections for left conjugate movement of both eyes through the medial longitudinal bundle, and nuclei of third, fourth, and sixth cranial nerves of same and opposite sides

contralateral relative afferent pupillary defects. It is important to remember that the tract and the lateral geniculate nucleus have a dual blood supply; therefore primary vascular lesions are uncommon. Most cases are caused by trauma, tumors, or arteriovenous malformations.

The inferior portion of the optic radiation passes through the temporal lobe and the superior portion passes through the parietal lobe. An insidious onset with mild neurologic deficits would be more typically neoplastic, whereas an acute event would be vascular.

Vascular lesions of the occipital lobe are common and account for over 80 % of cases of isolated homonymous visual field loss in patients over age 50. Because of the frequent presence of a dual blood supply, vascular occlusions may selectively spare the posterior cortex and produce homonymous defects with macular sparing (see lesion no. 5 in Fig. 8.19).

8.29 Neuromuscular Lesions—Supranuclear

A lesion of the brainstem will cause gaze palsies of conjugate movements and after vascular accidents will result in arteriovenous malformation, multiple sclerosis.

Fig. 8.19 Course of visual and pupillary fibers through the visual pathway and field of vision defects for lesions at sites 1, 2, 3, 4, and 5

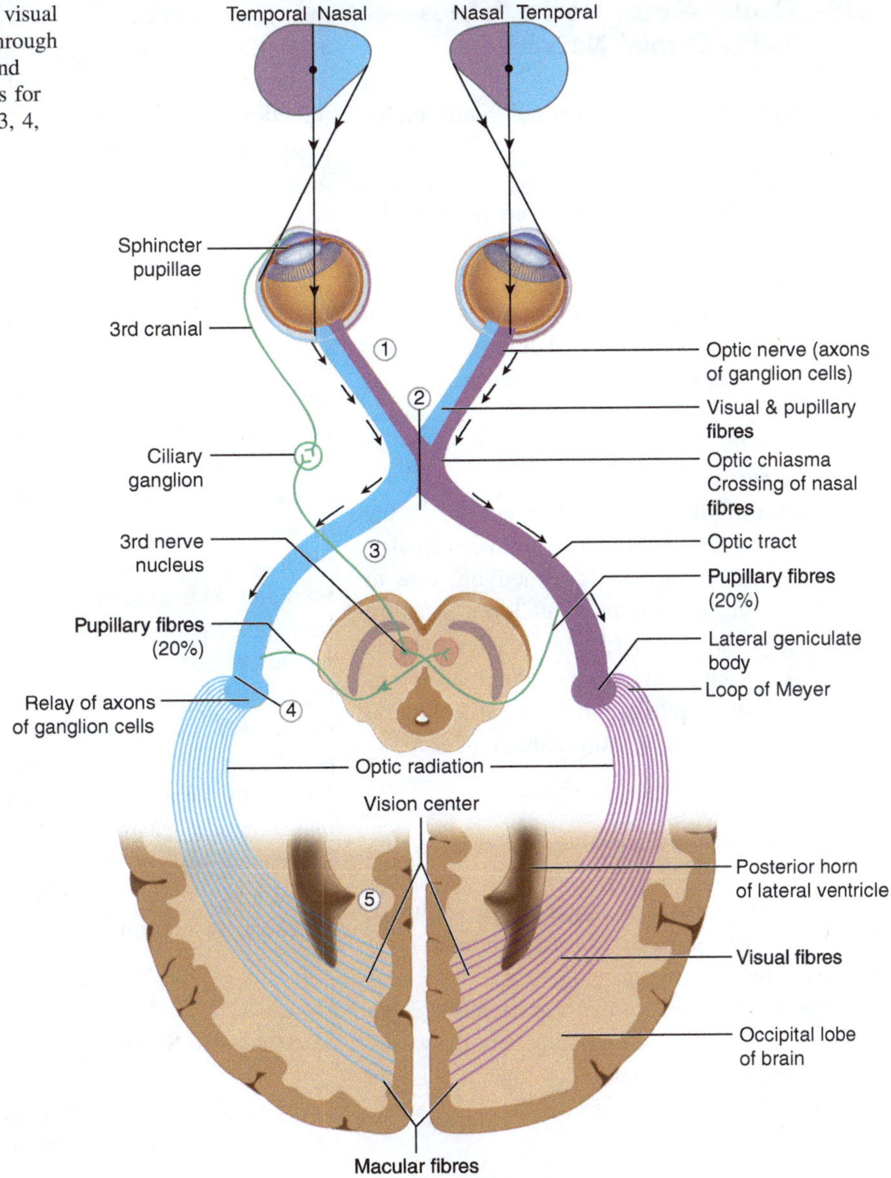

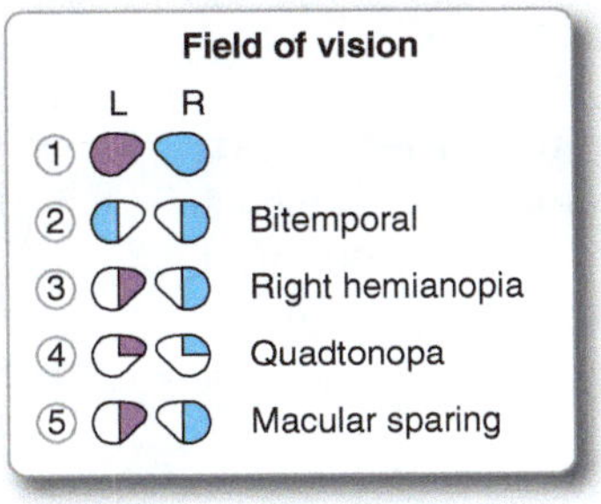

8.30 Ocular Motor Nerve Palsies—Third Cranial Nerve

Third nerve palsy in adults is usually attributable to ischemia (diabetes, hypertension, hyperlipidemia, and systemic vasculitis). Intracranial tumors may cause third cranial nerve palsy by direct damage to the nerve or as a result of mass effects. Bilateral peripheral third nerve palsy can be caused by interpeduncular lesions such as a basilar artery aneurysm. In ischemic lesions the pupils are not affected, but compression lesions by an aneurysm cause complete pupillary paralysis because the pupillary fibers in the optic tract are on the surface.

From the midbrain to its termination in the orbit, third cranial nerve palsy produces ipsilateral dysfunction; the eye is turned out due to intact lateral rectus muscle. In children the third nerve palsy may be attributable to migraine, meningitis, or be postviral or congenital. Symptoms are bilateral ptosis and bilateral limitation of elevation, depression, and adduction.

8.31 Trochlear Palsy

Trochlear palsy may be congenital or acquired. The acquired disorder is mostly traumatic. The nerve is vulnerable to injury at the site of the exit from the dorsal aspect of the brainstem. The eye is in upward deviation and hypertropic, which increases when the patient looks down and to the opposite side.

8.32 Sixth Nerve Palsy

In this condition the nerve runs a long course over the tip of the petrous portion of the temporal bone into the cavernous sinus. It has a 90° bend, and its palsy does not have much localizing value. This is the most common isolated single muscle extraocular palsy. Ischemia is a common cause (i.e., diabetes, migraine, and hypertension). The palsy may be involved in a middle ear infection. A child with an isolated sixth nerve palsy should be evaluated for a brainstem tumor or a glioma.

8.33 Migraine

Migraine headaches are a common recurrent illness of unknown cause. They usually manifest between 15 and 30 years of age, are more common in females with a family history, and are characterized by severe unilateral headache with or without an aura with flashes of light, dark, and colored spots in front of the eyes, nausea, and vomiting. Many factors including emotional ones may predispose or contribute to attacks. The whole episode may last 15–30 min and may be followed by a homonymous hemianopia on the nerve side lasting for several hours. Rarely, permanent visual loss may occur, which may be caused by cerebral infarction or arteriovenous malformation. Migraine sufferers may also experience episodes of transient monocular visual loss, thought to be due either to retinal or choroidal vasospasm.

Suggested Reading

Parsons' Diseases of the Eye, edn 16. Revised by Miller SJH. London: Churchill Livingstone; 1979. pp. 1–14; 545–60

Riordon-Eva P, Cunningham E (2011) Vaughan and Asbury's general ophthalmology, edn 18. McGraw Hill, New York, pp 1–26

Trevor-Roper PD (1974) The eye and its disorders. Blackwell Scientific Publications, London, pp 26–44, 323–34

Wolff's E (1976) Anatomy of the eye and orbit, edn 7. Revised by Warwick R. London: HK Lewis; 1976; pp. 1–29; 30–180D

9 Congenital Anomalies of Eye

Keywords

Congenital anomalies of eye · Ptosis · Coloboma of lid · Iris · Choroid · Lens and optic disc · Leukocoria · Retinoblastoma · Retinopathy of prematurity · Persistent hyper plastic primary vitreous · Albinism · Binocular vision · Polycoria · Aniridia · Subluxation of lens · Congenital cataract · Congenital glaucoma · Interstitial keratitis · Marfan syndrome · Simultaneous almost similar bifoveal image · Vision deprivation · Amblyopia · Anophthalmos · Cryptophthalmos

Careful examination of infant eyes soon after birth may reveal congenital anomalies such as ptosis, opacities in transparent structures, a lack of parallelism, albinism, and/or an uncorrected, high refractive error in one eye that may interfere with the normal development of binocular vision, which may continue until nearly age 10 depending on the simultaneous, nearly identical bifoveal image formation in both eyes. Because these conditions need timely, proper management by an ophthalmologist, early detection is mandatory so the eyes do not suffer from vision deprivation, leading to amblyopia and poor vision.

As part of the newborn physical examination, all infants should have their eyes examined. The tools needed for this are a good hand light, a direct ophthalmoscope, a binocular loupe for magnification, and a portable slit lamp.

In general, one must be sure to look for:

- The presence of a normal red reflex in pupillary area in both eyes;
- Normal external ocular anatomy;
- Parallelism by watching a central corneal reflex in two eyes at 6 months old;
- The ability to fixate light and follow a target (e.g., a small, squeaking, colored-plastic animal).

A careful eye examination soon after birth may reveal congenital abnormalities that suggest the presence of abnormalities elsewhere in the body. Congenital ocular abnormalities fall into two groups:

1. Developmental anomalies with genetic defects;
2. Tissue reactions to intrauterine insults (e.g., infections or drugs).

One simple approach should be that all newborns with white pupil (leukocoria) and history of prematurity are immediately referred to an ophthalmologist for possible retinoblastoma, hyperplastic primary vitreous, and premature retinopathy.

M.W. Ansari and A. Nadeem, *Atlas of Ocular Anatomy*, DOI 10.1007/978-3-319-42781-2_9

9.1 Globe Abnormalities

1. Anophthalmos (absence of eye—due to formation failure of optic vesicle)
2. Cystic eye due to invagination failure
3. Cryptophthalmos due to failure of eyelids to separate
4. Abnormally small eyes can be nanophthalmos (normal function) and microophthalmos (abnormal function).

9.2 Lid Abnormalities

1. Congenital ptosis (drooping of lid)—if unilateral can cause visual deprivation leading to amblyopia)
2. Coloboma (gap in upper or lower lid due to incomplete fusion of fetal maxillary processes)—may be associated with colobomas in iris, choroid, and optic disc.

Corneal abnormalities like opacities may be due to most common cause, congenital glaucoma, in which eyes are larger than normal (buphthalmos) and cornea also becomes larger. Megalocornea (larger cornea) may exist with normal function. All these cases will be referred to an ophthalmologist. Corneal opacities due to forceps delivery will clear in time. Interstitial keratitis is seen in congenital syphilis.

9.3 Iris and Pupillary Defects

Incomplete closure of the fetal cleft can give rise to colobomas (gaps) in iris, choroid lens, and optic nerve. Polycoria (multiple pupils) or aniridia (absence of iris) are rare. Albinism is due to absence of pigments in ocular structures and is associated with poor vision, nystagmus, and squint.

9.4 Lens Abnormalities

Congenital cataract (opaque lens) is the most common lens abnormality. Subluxation (partial displacement) is seen in Marfan syndrome. Maternal rubella during first trimester of pregnancy is a common cause in developing countries; if not treated early, may cause amblyopia.

9.5 Dermoids

Congenital rests of surface ectodermal tissues may lead to the formation of extraocular growths (dermoids), mostly superomedially at the limbus.

9.6 Delayed Canalization of Nasolacrimal Duct

Canalization of the nasolacrimal duct (NLD) normally occurs before birth or during the first month of life. Thirty percent of infants may have watering of the eye (epiphora) that can be easily cured by massaging just below the inner canthus and applying erythromycin eye drops 3 times a day for 3 months. In rare cases, epiphora may persist and lead to chronic dacryocystitis, which must be referred to an ophthalmologist for probing or surgery. A blind baby with a normal ocular and neurological examination history should be urgently referred to an ophthalmologist to prevent amblyopia. Normal development of the visual system depends on a simultaneous, nearly identical bifoveal image on the retina, for which eyes should be parallel, and free from squint, high refractive error, and opacities in the cornea and lens. If any of these conditions occur, they will interfere with normal development of binocular vision. The structural and functional development of macula is completed by the age 3 (visual acuity, 20/20).

Suggested Reading

Parsons' Diseases of the Eye, edn 16. Revised by Miller SJH. London: Churchill Livingstone; 1979. pp. 1–14; 545–60

Riordon-Eva P, Cunningham E (2011) Vaughan and Asbury's general ophthalmology, edn 18. McGraw Hill, New York, pp 1–26

Trevor-Roper PD (1974) The eye and its disorders. Blackwell Scientific Publications, London, pp 26–44, 323–34

Wolff's E (1976) Anatomy of the eye and orbit, edn 7. Revised by Warwick R. London: HK Lewis; 1976; pp. 1–29; 30–180D

Index

Note: Page number followed by f and t indicate figures and tables respectively

M.W. Ansari and A. Nadeem, *Atlas of Ocular Anatomy*, DOI 10.1007/978-3-319-42781-2

FSC
www.fsc.org
MIX
Papier | Fördert gute Waldnutzung
FSC® C083411

Kolibri 360 GmbH
Ferdinand-Jühlke-Straße 7
99095 Erfurt, Deutschland
produktsicherheit@kolibri360.de